GW00689623

NEPAL'S FOREST BIRDS:
THEIR STATUS AND CONSERVATION

by

Carol Inskipp

ICBP Monograph No. 4

Copyright © 1989 International Council for Bird Preservation.
32 Cambridge Road, Girton, Cambridge CB3 0PJ, U.K.

All rights reserved

British Library Cataloguing in Publication Data

Inskipp, Carol
    Nepal's Forest Birds: Their Status and Conservation
    1. Nepal. Birds. Conservation
    I. Title II. International Council for Bird Preservation
    III. Series
    639.9'78'095496

ISBN   0-946888-16-7
ISSN   1012-6201

Printed and bound by Labute Ltd, Cambridge.

Cover photograph: The Modi Khola valley lies below the spectacular Himalayan peak, Machhapuchhre (the 'Fish tail'), in the proposed Annapurna Conservation Area and is one of Nepal's species-rich forest areas. (Photo T. Inskipp)

INTERNATIONAL COUNCIL FOR BIRD PRESERVATION

ICBP is the longest-established worldwide conservation organisation. Its primary aim is the protection of wild birds and their habitats as a contribution to the preservation of biological diversity. Founded in 1922, it is a federation of 330 member organisations in 100 countries. These organisations represent a total of over ten million members all over the world.

Central to the successful execution of ICBP's mission is its global network of scientists and conservationists specialising in bird protection. This network enables it to gather and disseminate information, identify and enact priority projects, and promote and implement conservation measures. Today, ICBP's Conservation Programme includes some 100 projects throughout the world.

Birds are important indicators of a country's environmental health. ICBP provides expert advice to governments on bird conservation matters, management of nature reserves, and such issues as the control of trade in endangered species. Through interventions to governments on behalf of conservation issues ICBP can mobilise and bring to bear the force of international scientific and popular opinion at the highest levels. Conferences and symposia by its specialist groups help to attract worldwide attention to the plight of endangered birds.

ICBP maintains a comprehensive databank concerning the status of all the world's threatened birds and their habitats, from which the Bird Red Data Books are prepared. A series of Technical Publications gives up-to-date and in-depth treatment to major bird conservation issues. This series of Monographs (of which the present volume is the third) provides comprehensive, up-to-date information on specific or regional issues relating to bird conservation.

ICBP, 32 Cambridge Road, Girton, Cambridge CB3 0PJ, U.K.

U.K. Charity No. 286211

Blood Pheasant *Ithaginis cruentus* (Dave Showler)

In view of the ecological spectrum and the diversities in wildlife resources of Nepal we share a common concern with all those who realise the need to conserve the precious heritage of world fauna.

November 22 1979

In our view, birds show the perennial beauty of creation and together with other fauna and their landscape, they make up the natural heritage of a nation.

October 29 1979

His Majesty Birendra Bir Bikram Shah Dev,
King of Nepal

CONTENTS

FIGURES

ACKNOWLEDGEMENTS

I am indebted to the International Council for Bird Preservation British Section for funding this study and resulting publication.

I am grateful to the Department of National Parks and Wildlife Conservation and the King Mahendra Trust for Nature Conservation in Nepal for their support and warmly thank Hemanta Mishra, King Mahendra Trust for Nature Conservation, Tirtha Bahadur Shrestha, Royal Nepal Academy, Michael Green, Protected Areas Data Unit, IUCN and Adam Gretton, International Council for Bird Preservation for their useful comments on the text.

The study would not have been possible without the generous assistance of many people who sent me their bird records and habitat information. Special thanks go to Jack Cox Jr, Hari Sharan Nepali and Arend van Riessen for their important contributions. In addition, Jack advocated the need for the protection of a representative area of tropical evergreen forest long before this study was started. I am also most grateful to: Per Alind, Per Alström, Tim Andrews, Carl Axel-Bauer, Tony Baker, Mark Beaman, Jan Bolding, Paul Bradbear, Dick Byrne, Steen Christensen, Andy Clements, David Clugston, Bernard Couronne, Ian Dawson, Frank de Roder, Philippe Dubois, Nick Dymond, Jon Eames, Pete Ewins, Vernon Eve, Richard Fairbank, Dave Farrow, Robert Fleming Jr, Steve Gantlett, Simon Gawn, Andrew Goodwin, Michael Green, Richard Grimmett, Kaj Halberg, Jim Hall, Phil Hall, John Halliday, Patrick Hamon, Simon Harrap, Andrew Harrop, Phil Heath, Joel Heinen, Pete Hines, Goran Holmström, Tim Inskipp, Rob Innes, Paul Jepson, Ron Johns, Bas Jongeling, Torben Jorgensen, Rafi Juliusburger, Stan Justice, Pete Kennerley, Ben King, Nils Kjellen, Jaroslav Klapste, Jean-Christophe Kovacs, Erling Krabbe, Frank Lambert, Tony Lelliott, Paul Lewis, Vaughan Lister, Steve Madge, Sjoerd Mayer, Jochen Martens, Rod Martins, David Mills, Hiroyuki Morioka, Chris Murphy, Kathleen Munthe, Serge Nicolle, Herbet Nickel, Thomas Nilsson, Mike Parr, Ib Petersen, Oleg Polunin, Richard Porter, Anders Priemé, Nick Preston, David Pritchard, Nigel Redman, Tim Reid, Gerry and Lucy Richards, Jimmy Roberts, Robert Roberts, Philip Robinson, Tim Robinson, Steve Rooke, Craig Robson, Jonathan Ross, John Rossetti, Valentine Russell, Jelle Scharringa, Mike Searle, Peter Sieurin, Neil Simpson, Stewart Smith, Werner Suter, Jean-Marc Thiollay, Dave Thorns, Mick Turton, Arnoud van den Berg, Göran Walinder, Jill Warwick, naturalists of West Nepal Adventures (P) Ltd, James Wolstencroft and Mark Wotham.

Thanks go to the following people for providing invaluable information or advice on forestry: T. M. Abell, Land Resources Development Centre, Gina Green and Tony Greaves, Oxford Forestry Institute, and especially to John Hudson, Team Leader Nepal-United Kingdom Forestry Research Project. I am grateful for the facilities provided by the Oxford Forestry Institute Library and the help given by the librarian Jasmine Howse.

I thank Paul Goriup, former Project Director for help in the early stages, Sue Wells, Assistant Director (Information) for organising this monograph's publication and to Gina Pfaff, all from ICBP, for typing the manuscript. I am particularly grateful to Tony Diamond for assistance and Mike Rands, the present Project Director of ICBP and my husband Tim for their encouragement, help and useful criticism.

Fire-tailed Myzornis *Myzornis pyrrhoura* (Dave Showler)

SUMMARY

Nepal is of great value for birds mainly because of its wide variety of forests which comprise tropical, subtropical, temperate, subalpine and alpine types. The high proportion of 77 percent of Nepal's breeding birds and 93 percent of those for which the country may hold significant world populations utilise forests or shrubs.

The protected area coverage and threats to all Nepal's breeding birds are assessed. There are just 29 of all breeding birds which are currently recorded only outside the protected and proposed protected areas in the nesting season. However, a large number of breeding birds (22 percent of the total) are considered to be at risk and 84 percent of these are dependent on forests.

Forest losses and deterioration are by far the greatest threats to Nepal's birds. Forest resources are declining chiefly because they can no longer meet the requirements of the population. The vast majority of Nepalis depend on forests for their essential requirements for fuel, animal fodder and other basic materials. Conservation of Nepal's forests is therefore vital for the future of its people as well as its birds.

A combination of deforestation and over-grazing has caused massive soil erosion and the rapid run-off of rain during the monsoon. The resulting widespread flooding in the lowlands has led to enormous loss of life, crops and property. Whilst there has recently been a great expansion in afforestation, the overall impact has been very small. The new plantations of indigenous broad-leaves are more beneficial to birds than conifers. Improved management of existing low density forests would be much more valuable however and there is great potential for this aspect of forestry.

Only 16 percent of all forest birds have adapted to breed in habitats heavily modified or created by people. A few species which prefer open forests or scrub must have increased as a result of forest depletion, but nearly all of them are common and widespread. Overall the populations of most Nepalese forest birds are likely to have decreased.

The status of Nepal's forest birds is assessed and the breeding bird communities within the country's main different forest types identified. Subalpine and upper temperate forests are the most internationally important for breeding birds as they support high numbers of species which may have significant world populations in Nepal (58 and 65 respectively). Lower temperate forests, especially the wetter types, are also internationally important as they hold 42 breeding birds which may have significant populations in the country. Tropical and subtropical forests have the highest diversity of breeding species and also the greatest numbers of species considered to be at risk in Nepal (55 and 42 respectively).

The Kali Gandaki valley which runs north/south through almost the middle of Nepal is an important divide for bird distribution. Western forests are poorly recorded compared to those in the east, but when all likely unrecorded species are counted, eastern forests are still considerably richer. They also support a much larger number of species for which Nepal may hold significant world populations.

Species inventories annotated with status are compiled for Nepal's protected and proposed protected areas and their importance for birds described. The two proposed areas, the Annapurna Conservation Area and the Barun valley together with Langtang National Park, are the most internationally important of Nepal's protected areas for birds. The Royal Chitwan, Sagarmatha and Langtang National Parks and the Annapurna Conservation Area are well studied, but the others are under-recorded.

The protected area coverage of some Nepal's forests, which are particularly rich in bird species, is summarised. These forests are also likely to support a high diversity of other groups of fauna and flora.

An account is given of the two main unprotected species-rich forests – Phulchowki mountain in the Kathmandu valley and the Mai valley in the far east. Their importance for birds and threats are described and species inventories annotated with status included.

The representation of Nepal's forest types by the current protected and proposed protected areas is reviewed. The great majority, including all upper temperate, subalpine, alpine and most tropical forest types, are already well represented.

Protection of forests of the three main unrepresented types is urgently recommended. These are tropical evergreen forests, subtropical and lower temperate broadleaved forests in the far east (Mai valley area) and subtropical broadleaved forests further west. The most outstanding gap in representation is of subtropical broadleaved forests as they once covered much of central and east Nepal. Protection of Phulchowki's forests is highly recommended as they comprise a good example of the above-mentioned forest type as well as temperate forests important for their fauna and flora.

Survey work is needed in remaining forests of the unrepresented types to identify new areas suitable for protection as soon as possible. The location of these forests which were in reasonable condition (i.e. with more than 40 percent tree crown cover) in 1978/79 are described and mapped using vegetational and the Land Resources Mapping Project land utilisation maps.

More fieldwork is required in western Nepal forests (west of the Kali-Gandaki river) as they are under-recorded.

More bird surveys are recommended in all the under-recorded protected areas and especially in the species-rich Barun valley – a proposed protected area. This will be useful for their better management and will enable the identification of bird species still requiring habitat protection to be re-assessed.

A high proportion (36 percent) of Nepal's breeding species are altitudinal migrants which spend much of their time outside the nesting season at lower altitudes. Nepal's forests are also important for winter visitors from further north. A study of forests important for birds outside the breeding season is therefore needed.

Yellow-billed Blue Magpie *Urocissa flavirostris* (Dave Showler)

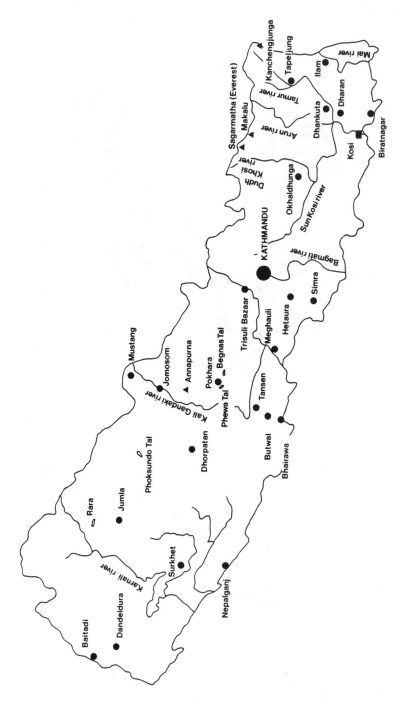

Figure 1: Nepal's major geographical features and places.

INTRODUCTION

The avifauna of Nepal is exceptionally diverse. About 850 bird species have been recorded, including 609 which breed or probably breed[1]. There are 124 species whose breeding distributions are restricted to an area encompassing the Himalaya, north-east India, northern south-east Asia and south-west China, for which Nepal may hold internationally significant populations[2]. The country may be especially important for 35 of these species. They either have particularly restricted ranges within the general area considered or have been described as uncommon or rare in the Indian subcontinent (Inskipp and Inskipp 1986)[3].

Nepal's species richness can be partly attributed to the wide range of altitude, climate and vegetation in the country. Forest types include wet tropical forests in the lowlands and dry coniferous subalpine forests on the northern faces of the Himalaya. The other major factor contributing to Nepal's species variety is its position of overlap between the Oriental realm to the south and the Palearctic realm to the north.

The country is of high value for birds mainly because of its rich forests. About 77 percent of Nepal's breeding bird species, 67 percent of wintering species and 93 percent of those for which Nepal may hold internationally significant breeding populations utilise forests and shrubs.

Once Nepal was extensively forested, but by 1979 only 42.8 percent of the country was forest land (that is covered by trees and shrubs) and much of this was in poor condition, with only a scattering of trees (Kenting 1986)[4]. Forests are becoming depleted at an increasing rate and are now gravely threatened. The main causes are rapid population growth and the dependence of the vast majority of people on forests for their basic needs. Forests provide as much as 87 percent of the nation's energy requirements (HMG/IUCN 1983), as well as fodder and bedding material for livestock and timber for construction. A combination of deforestation and over-grazing has caused massive soil erosion and the rapid run-off of rain during the monsoon. The resulting widespread flooding in the lowlands has led to great loss of life, crops and property.

Please note that throughout the report

[1] 'breeding species' includes both proved breeding and probable breeding species;
[2] species described as having 'significant breeding populations in Nepal' refer to those with internationally significant breeding populations in the country as defined above;
[3] this is the definition of 'species for which the country may be especially important';
[4] the term 'forests' refers to both forests and shrubs.

In the face of these threats the Department of National Parks and Wildlife Conservation in Nepal is aiming to protect a representative sample of the country's ecosystems. In April 1987 protected areas covered 7.4 percent of the country, comprising six national parks, four wildlife reserves and one hunting reserve (Oy and Madecor 1987a). Although extensive, the network is far from complete and new protected areas are needed to ensure the future of all ecosystem types.

This report aims to:

1. Identify the chief breeding bird communities within Nepal's different forest types.

2. Assess how well the forest types are represented by the current protected areas and proposed protected areas.

3. Identify forest types important for bird conservation which are so far unprotected or have limited protection. Given that faunal diversity is directly related to floral diversity, the results can be used to identify forests which are valuable for other wildlife as well as birds.

4. Compile bird species inventories annotated with status for all protected areas and proposed protected areas, and to identify breeding bird species only recorded outside of them.

5. Compile a list of all breeding bird species considered to be at risk in Nepal.

6. Summarise present conservation measures, including afforestation schemes.

7. Assess the status of, and threats to forest species.

8. Make recommendations for future survey work and the establishment of additional protected areas.

TAXONOMIC NOTE

Bird names follow Inskipp and Inskipp (1985) apart from one recent taxonomic change. Collared Scops Owl *Otus lempiji* and Indian Scops Owl *O. bakkamoena* are now both considered to occur (King and Roberts 1986).

CLIMATE

Climate and geography are briefly discussed first as they are major factors determining the occurrence and distribution of the different forest types.

Nepal's climate is dominated by the monsoon. It is extremely varied ranging from tropical in the lowlands to arctic in the high peaks (Table 1). Altitude and exposure to the sun and rain are the two most influential climatic factors. About 90 percent of rain falls during the monsoon between June and December. The moist air flowing north and west from the Bay of Bengal falls as rain on the southern slopes of the Himalaya. The districts north of the range - Mustang, Manang and Dolpo lie in the rain shadow and have very low precipitation. Western Nepal is much drier than the east as the monsoon rains reach there later and the monsoon lasts for a shorter period. Rainfall distribution depends on altitude as well as latitude. The amount of rainfall increases with altitude to a maximum and then decreases again (Table 2) (Singha 1984-1985). Aspect of slope greatly affects climate in the Himalaya. Southern slopes are considerable warmer and sunnier than those facing north, invariably resulting in quite different and much drier vegetation.

Table 1: Average temperatures (Centigrade) for some places in Nepal.

Place	Namche Bazar		Jomosom		Kathmandu		Pokhara		Meghauly (Terai)	
Altitude (m)	3440		2650		1300		800		180	
	Max	Min	Max	Min	Max	Min	Max	Min	Max	Min
Jan	6	6	11	2	17	2	19	6	23	8
Feb	6	4	12	1	21	3	22	8	28	11
Mar	8	2	16	2	25	9	27	14	33	17
Apr	11	1	20	4	27	11	31	17	37	19
May	14	3	23	7	28	16	31	19	37	22
Jun	15	7	25	13	28	19	31	21	34	23
Jul	16	8	25	14	28	20	30	21	33	25
Aug	16	8	25	14	27	20	29	21	32	24
Sept	15	7	23	11	26	19	28	20	32	24
Oct	11	1	13	5	25	12	27	15	31	19
Nov	8	3	15	1	21	8	24	11	27	14
Dec	7	4	13	2	19	3	20	8	24	8

Source: Metereology Dept (HMG)

Table 2: Variation of annual rainfall with altitude and latitude.

Altitude	Rainfall (mm)		
(m)	West	Central	East
200	1,120	1,900	1575
400	1,200	2,100	1688
600	1,280	2,300	1800
800	1,360	3,000	1900
1,000	1,440	2,300	2000
1,200	1,520	2,025	2100
1,400	1,400	1,725	2175
1,600	1,150	1,500	2215
1,800	975	1,325	2250
2,000		1,125	2275
2,400		712	2036
2,800		275	1825
3,200			225

Source: Singha (1984-1985)

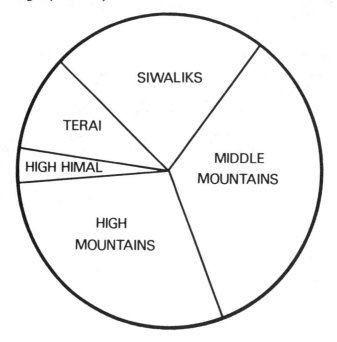

Source: Kenting 1986

Figure 2: Percentage of forest areas in the main physiographic zones.

GEOGRAPHY AND VEGETATION

The Kingdom of Nepal is remarkable for its rugged topography and diverse vegetation. The altitude ranges from 70 m above sea level to 8848m at the summit of Sagarmatha (Mount Everest), the highest point on earth. The country is landlocked between China to the north and India on the east, west and south. It lies between the latitudes of 26°20'N and 30°26'N and between the longitudes of 80°15'E and 88°10'E. Most of Nepal lies in the Himalaya and forms the central part of the range, one third of its entire length. The country's area is 147,484 sq km, little bigger than England and Wales combined and averages about 870 km from east to west.

It can be divided into five major physiographic zones which run roughly parallel from north-west to south-east:

    High Himal
    High Mountains
    Middle Mountains
    Siwaliks
    Terai

These are the zones used by the Land Resources Mapping Project which produced the first accurate measure of the coverage and status of Nepal's forests (Kenting 1986). As there is a wide variety of habitats within each of these zones Oy and Madecor (1987a) have suggested their extension to cover:

    Trans Himalayas
    High Himalayas
    Middle Mountains
    Middle Mountain Valleys
    Mahabharat
    Siwaliks
    Inner Terai (Bhabar)
    Outer Terai

However, as they pointed out, even these extended zones do not take into account all the vegetation differences in Nepal. There is an extraordinary diversity of forests in the country. Stainton (1972) has identified 35 different types. Several of them occur within each physiographic zone and often overlap into other zones. In this report the vegetation is classified by altitude and climatic zone (tropical, subtropical, lower temperate, upper temperate, subalpine and alpine) following Dobremez (1976) and summarised by Jackson (1987) see p.43. Certain types described by Stainton and Dobremez which only occur in very restricted areas have either been omitted following Jackson (1987) or treated together with other forest types. These are *Castanopsis tribuloides-C. hystrix, Lithocarpus pachyphylla, Aesculus-Juglans-Acer, Tsuga dumosa, Picea smithiana, Abies pindrow, Cedrus deodara, Cupressus torulosa, Larix* and *Populus ciliata* forests and *Hippophae* scrub. Although available bird records for these forest types are very limited, they are undoubtedly important for some species, for hemlock *Tsuga dumosa*. Further study of them is needed, at least of their birds and almost certainly of other groups of fauna and flora. Forests of mixed types are very important and indeed some bird species are mainly found in them, for example, the Spotted Laughing-thrush *Garrulax ocellatus* which frequents mixed conifers, maples, rhododendrons and bamboo (Fleming 1984 *et al.*). Mixed conifer/hardwood forests comprised 21 percent of all forest land in 1979 (Kenting 1986).

The forest land areas and percentages listed below are taken from Kenting (1986) and are based on data collected in 1978/79.

TERAI

Altitude: 70-330 m

Climatic zone: Tropical

Forest types: Shorea robusta, Acacia catechu-Dalbergia sissoo, tropical evergreen forest.

Forest land area: 5,913 sq km

Percentage of total forest land in Nepal: 9.5 percent
The terai is a narrow lowland strip lying north of the Indian border. It is a continuation of the Gangetic Plain and contains the most productive agricultural land in Nepal. It also includes the dry bhabar region, consisting of gravelly soil which has been washed down from the foothills and accumulated at their base. The bhabar is not very suitable for agriculture and large tracts of forest remain here.

SIWALIKS

Altitude: 200-1,500 m

Climatic zones: Tropical, subtropical

Forest types: West Nepal: *Shorea robusta, Pinus roxburghii* and *Terminalia-Anogeissus* deciduous hill forests.

East Nepal: *Shorea robusta, Schima-Castanopsis, Terminalia-Anogeissus* deciduous hill and subtropical evergreen forests.

Forest land area: 14,447 sq km

Percentage of total forest land in Nepal: 23.4 percent
The Siwaliks, the first of the Himalayan foothills, rise steeply north of the bhabar region. Much of the forest on the hills remains uncut because the soils are shallow and erodable and have little potential for cultivation. In central and west Nepal, the Siwaliks are separated from the next range of mountains to the north, the Mahabharat range, by broad gently sloping valleys known as duns. The duns include the valleys of large rivers, such as the Karnali, Babai and Rapti. Some of the dun valleys are now well cultivated.

MIDDLE MOUNTAINS

Altitude: 800-2,400 m

Climatic zones: Subtropical, lower temperate

Forest types: West Nepal: *Pinus roxburghii, Alnus nepalensis, Quercus leucotrichophora, Q. lanata* and *Q. floribunda, Pinus wallichiana.*

East Nepal (wetter part) and south of Annapurna and Himal Chuli: *Schima-Castanopsis*, riverine forest with *Toona* and *Albizia*, *Alnus nepalensis*, lower temperate mixed broadleaves and *Quercus lamellosa*.

Forest land area: 17,941 sq km

Percentage of total forest land in Nepal: 35 percent
The Middle Mountains region is a broad complex of mountains and valleys, including the Mahabharat range which lies north of the Siwaliks and runs from WNW to ESE across the whole of south Nepal. River valleys in the Middle Mountains are often very deep and narrow with steep slopes rising to 2,000 m or more. According to the 1981 census, 46 percent of the population lived in the region and nearly a quarter of a million people in the capital city, Kathmandu. Overpopulation, over-grazing and forest degradation are major problems here. Little forest remains except in the less populated west. In the centre and east, especially between 1,000 m and 2,000 m, nearly all the natural forest has been replaced by intensive terraced cultivation. Often the only woody vegetation left is in the form of 'shrubberies' (a term used by Stainton (1972) to describe dense masses of shrubs up to about 3 m in height). The region is less cultivated above 2,000 m.

HIGH MOUNTAINS

Altitude: 2,200-4,000 m

Climatic zones: Lower temperate, upper temperate, subalpine

Forest types: West Nepal: *Quercus leucotrichophora* and *Q. lanata*, *Pinus wallichiana*, *Q. semecarpifolia*, *Juniperus indica*, *Abies spectabilis*, *Betula utilis*.
East Nepal and south of Annapurna and Himal Chuli: *Quercus lamellosa*, upper temperate mixed broadleaved, *Rhododendron* spp., *Abies spectabilis*, *Betula utilis*.

Forest land area: 16,315 sq km

Percentage of total forest land in Nepal: 28.6 percent
Although steep slopes of some of the high mountains are intensively terraced, the region is generally less cultivated and less populated than the Middle Mountains. Large scale cultivation ceases above 2,400 m. The forests are in a better condition than elsewhere in Nepal and include some of the least disturbed forests, especially those of conifers, such as *Abies spectabilis*. This zone also includes part of the trans-Himalayan zone, which lies north of the range and has considerably less rainfall than the rest of Nepal. Much of it is cold semi-desert with a climate and vegetation similar to the Tibetan steppe to the north.

HIGH HIMAL

Altitude: 4,000 m+

Climatic zone: Alpine

Forest types: *Rhododendron* spp., *Juniperus* spp.

Forest land area: 1,552 sq km

Percentage of total forest land in Nepal: 3.5 percent
Much of the region is under permanent ice and snow. There is a little cultivation in the valleys, but most land is only suitable for summer grazing. The region's spectacular mountains include all the world's ten peaks exceeding 8,000 m and

these are attracting increasing numbers of trekkers and mountaineers every year. The High Himal also contains some of the trans-Himalayan zone.

Whiskered Yuhina *Yuhina flavicollis* (Linda Schrijver)

CONSERVATION MEASURES

LEGISLATION

A brief summary of conservation legislation is given below. Detailed information is presented by Green (1986) and Oy and Madecor (1987a).

Nepal's conservation programme was initiated by His Majesty's Government in 1971. The basis for protected areas administration and wildlife conservation was provided by the National Parks and Wildlife Conservation Act 2029 introduced in 1973. Nature conservation is the responsibility of the Department of National Parks and Wildlife Conservation (DNPWC), a sector of the Ministry of Forests and Soil Conservation. The department administers parks, reserves and Kathmandu Zoo and coordinates the activities of wildlife guards. The Royal Nepal Army is responsible for the provision of guards and law enforcement in the protected areas.

All forests were nationalised in 1957 in an attempt to protect them, but unfortunately in many areas forest depletion accelerated instead.

In 1978 the Panchayat Forests and Panchayat Protected Forests regulations were introduced enabling the Forest Department to return ownership and control of forest land to local communities, so encouraging them to conserve forest resources and plant trees on unproductive land.

Watershed protection areas are managed by the Department of Soil Conservation and Watershed Management (HMG/IUCN, 1983). The National Parks and Wildlife Conservation Regulations of 1974 control hunting and restrict trade in wild animals in accordance with the Convention on International Trade in Endangered Species of Wild Fauna and Flora (CITES).

A National Parks and Wildlife Development Plan is being prepared and a draft was available in 1987 (Oy and Madecor 1987a). The plan's objectives are to identify constraints and issues hampering smooth development of the DNPWC and to outline a systematic scheme for rational development of protected areas in Nepal.

The King Mahendra Trust for Nature Conservation (KMTNC), established in 1982, is a non-governmental, non-profit making and independent organisation. The Trust aims to conserve and manage natural resources in order to improve the quality of life of the Nepalese people. It assists and complements the efforts of the Nepal government and foreign agencies engaged in the field of nature conservation (Rana *et al.* 1984).

The International Centre for Integrated Mountain Development is based in Kathmandu. The centre's primary objective is to promote economically and environmentally sound development in the Hindu-Kush and Himalaya and to improve the well being of the local populations.

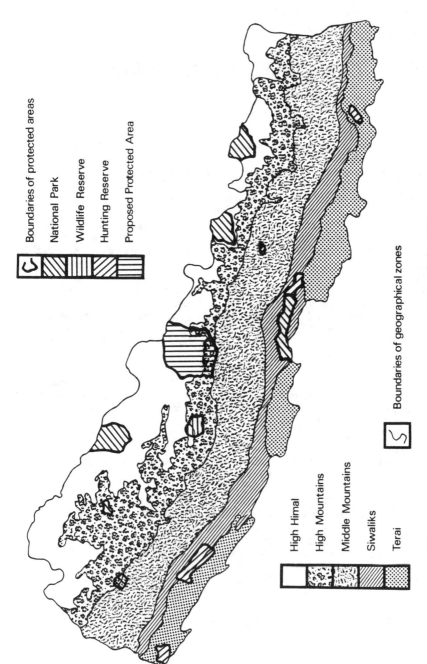

Boundaries of protected areas

National Park

Wildlife Reserve

Hunting Reserve

Proposed Protected Area

Boundaries of geographical zones

High Himal

High Mountains

Middle Mountains

Siwaliks

Terai

Figure 3: Nepal's protected areas and main physiographic zones.

PROTECTED AREAS

Protected area categories

Legislation makes provision for four categories of protected areas (Oy and Madecor 1987a):

National parks – areas set aside for conservation, management and utilisation of animals, birds, vegetation or landscape together with the natural environment.

Strict nature reserves – areas of ecological significance or other significance set aside for scientific study (none has so far been designated).

Wildlife reserves – areas set aside for the conservation of animal and bird resources and their habitats.

Hunting reserves – areas set aside for the management of animal and bird resources for purposes of sport hunting.

By April 1987 six national parks, four wildlife reserves and one hunting reserve had been established (Oy and Madecor 1987a) and protected areas covered 7.4 percent of the country. In addition management of the Shivapuri Watershed and Wildlife Reserve will soon be the responsibility of the DNPWC.

There are proposals for two other protected areas. Protected status for the Annapurna region was first proposed in 1971. Conservation measures began in 1985 with the formation of the Annapurna Conservation Area Study project which is now operated by the KMTNC. The Barun Project is being carried out under the aegis of the DNPWC with funds provided by the Woodlands Institute, USA. The Institute is negotiating the eastwards extension of Sagarmatha National Park to include the Barun valley (Oy and Madecor 1987a). When these additional areas are designated, protected areas will cover over 10 percent of Nepal.

Table 3: Numbers of bird species in protected and proposed protected areas.

Column 1 Number of species recorded
Column 2 Number of breeding species
Column 3 Number of breeding species considered at risk in Nepal
Column 4 Number of breeding species which may have internationally significant populations in Nepal

Protected or proposed protected area	1	2	3	4
Chitwan	489	263	55	10
Khaptad	218	175	7	50
Langtang	283	246	17	84
Rara	153	87	1	32
Sagarmatha	152	103	0	36
Barun	169	159	25	66
Shey-Dolpo	105	96	0	18
Sukla Phanta	268	180	22	5
Bardia	256	193	22	9
Kosi Tappu	256	176	18	6
Shivapuri	149	100	6	26
Dhorpatan	136	124	4	41
Annapurna	441	329	38	100

White-browed Fulvetta *Alcippe vinipectus* (Dave Showler)

DESCRIPTION OF EXISTING AND PROPOSED PROTECTED AREAS

The following information is given below for each of Nepal's protected areas: an account of the importance for birds (all species), date gazetted, area, biogeographical province, physiographic zone(s), list of forest types and a brief description of geographical location. Bird species inventories annotated with status are given for each of the existing and proposed protected areas. Species breeding and wintering in the areas (but not passage migrants or vagrants), which may have internationally significant breeding populations in Nepal, and those considered at risk in the country, are indicated on the inventories. Numbers of bird species in protected and proposed protected areas are given in Table 3. Much more detailed information on aspects other than birds is given in Green (1986), including descriptions of geographical location, fauna, physical features, vegetation, cultural heritage, conservation management, scientific research and visitor facilities, and local population.

Royal Chitwan National Park (including Parsa Wildlife Reserve)
World Heritage Site

Dates gazetted: Royal Chitwan National Park 1973, Parsa Wildlife Reserve 1984

Areas: Royal Chitwan National Park 932 sq km, Parsa Wildlife Reserve 499 sq km

Location: Lies in south-central Nepal in a dun valley and extends into the Siwalik hills

Biogeographical province: Indus-Ganges Monsoon Forest

Physiographic zones: Siwaliks

Vegetation: Tropical forests: *Shorea robusta* (about 70 percent of the park), *Acacia catechu-Dalbergia sissoo*. Tropical evergreen (very small area of the park in gulleys and along streams): *Terminalia-Anogeissus* deciduous hill forest. Subtropical forest: *Pinus roxburghii*. Riverine grasslands (about 20 percent of the park).

Importance for birds (well-recorded): A species list is given on p.80 and consists of a comprehensive species list annotated with status compiled by Gurung (1983) with recent additions. A larger number of bird species has been recorded at Chitwan (489 in total) than any other protected area in Nepal. This can be attributed to the park's wide range of habitat types comprising forests, grasslands and wetlands and to its location. The park mainly lies in the tropical lowlands, the richest zone for bird species. As it is in the centre of the country species typical of both east and west Nepal occur. Chitwan supports a larger number of threatened Nepalese species[1] than any other protected area, 55 breeding species in total, 36 of which are endangered or vulnerable. There are ten breeding species for which Nepal may hold significant populations including the Bengal Florican *Houbaropsis bengalensis* and Rufous-necked Laughing-thrush *Garrulax ruficollis*. It is the only locality in the country for the latter species and also for Striped Buttonquail *Turnix sylvatica*, Bristled Grass Warbler *Chaetornis striatus* and

---

[1] Please note that throughout the report the terms 'threatened' and 'considered at risk' are used in the national context.

Slender-billed Babbler *Turdoides longirostris*. In addition Chitwan is the only protected area where the following threatened species have been found: Yellow Bittern *Ixobrychus sinensis*, Black Baza *Aviceda leuphotes*, Laggar *Falco jugger*, Blue-breasted Quail *Coturnix chinensis*, Thick-billed Green Pigeon *Treron curvirostra*, Mountain Imperial Pigeon *Ducula badia*, Vernal Hanging Parrot *Loriculus vernalis*, Red-winged Crested Cuckoo *Clamator coromandus*, Banded Bay Cuckoo *Cacomantis sonneratii*, Tawny Fish Owl *Ketupa flavipes*, White-vented Needletail *Hirundapus cochinchinensis*, Deep-blue Kingfisher *Alcedo meninting*, White-browed Piculet *Sasia ochracea*, Long-tailed Broadbill *Psarisomus dalhousiae*, Hooded Pitta *Pitta sordida*, White-throated Bulbul *Criniger flaveolus*, Lesser Necklaced Laughing-thrush *Garrulax monileger*, Greater Necklaced Laughing-thrush *G. pectoralis*, Ruby-cheeked Sunbird *Anthreptes singalensis* and Little Spiderhunter *Arachnothera longirostra*.

Chitwan is very important for wintering birds, both winter visitors from outside Nepal and many altitudinal migrants which descend to the lowlands outside the breeding season. A total of about 160 species, including over 20 wildfowl species, about 30 birds of prey and 16 waders are winter visitors. The park is a valuable staging point for numerous passage migrant species, including forest and water birds.

Langtang National Park

Date gazetted: 1976

Area: 1,710 sq km

Location: Lies in the central Himalayan region

Biogeographical province: Himalayan Highlands

Physiographic zone: High Himal

Vegetation: Tropical (0.2 percent of the park): *Shorea robusta*
Subtropical forests (2 percent of the park): *Schima wallichii/Castanopsis indica*, *Pinus roxburghii*. Lower temperate forests (4.8 percent of the park): *Quercus lanata*, *Q. lamellosa*, *Pinus wallichiana*. Upper temperate forests (9.9 percent of the park): *Quercus semecarpifolia* (often with *Tsuga dumosa*). Subalpine forests (21.5 percent of the park): *Abies spectabilis*, *Betula utilis*, *Tsuga dumosa*, *Larix* spp., *Rhododendron* spp., *Juniperus* spp., *Caragana* (has developed in the upper Langtang valley after removal of forests and over-grazing). Alpine zone (21.5 percent of the park): *Rhododendron* spp., *Juniperus* spp. Percentages of forest cover are taken from Green (1986).

Importance for birds (well-recorded): A species list is given on p.90. The park is one of the three most internationally important protected areas in Nepal for birds. It supports a rich variety of breeding species, 246 in total, including 84 species for which Nepal may hold significant populations. It is especially important for upper temperate and subalpine forest species. The park is the only place in Nepal where the Dark-rumped Rosefinch *Carpodacus edwardsii* has been recorded in the breeding season. Other breeding species include Satyr Tragopan *Tragopan satyra*, Ibisbill *Ibidorhyncha struthersii*, Orange-rumped Honeyguide *Indicator xanthonotus*, Bay Woodpecker *Blythipicus pyrrhotis*, Gould's Shortwing *Brachypteryx stellata*, Rufous-breasted Bush-Robin *Tarsiger hyperythrus*, Long-billed Thrush *Zoothera monticola*, Smoky Warbler *Phylloscopus fuligiventer*, Large Niltava *Niltava grandis*, Fulvous Parrotbill *Paradoxornis fulvifrons*, Scaly Laughing-thrush *Garrulax subunicolor*, Fire-tailed Myzornis *Myzornis pyrrhoura*,

Yellow-bellied Flowerpecker *Dicaeum melanoxanthum*, Vinaceous Rosefinch *Carpodacus vinaceus*, Crimson-browed Finch *Propyrrhula subhimachala*, Scarlet Finch *Haematospiza sipahi* and Spot-winged Grosbeak *Mycerobas melanozanthos*.

Sagarmatha (Mount Everest) National Park
World Heritage Site

Date gazetted: 1976

Area: 1,148 sq km

Biogeographical province: Himalayan Highlands

Physiographic zone: High Himal

Location: Situated in Solu-Khumbu district in north-east Nepal

Vegetation: Upper temperate: *Pinus wallichiana* Subalpine: *P. wallichiana, Abies spectabilis, Betula utilis.* Alpine zone: *Juniperus* spp.

Quercus semecarpifolia used to dominate upper temperate forests, but former stands of this species and of *Abies spectabilis* have been colonised by *Pinus wallichiana*.

Importance for birds (well-recorded): A species list is given on p.96. The park is important for a number of species breeding at high altitudes, such as Blood Pheasant *Ithaginis cruentus*, Robin Accentor *Prunella rubeculoides*, White-throated Redstart *Phoenicurus schisticeps*, Grandala *Grandala coelicolor* and several rosefinches. A total of 36 breeding species for which Nepal may have significant populations has been recorded. The park's small lakes, especially those at Gokyo are used as staging points for migration and at least 19 waterbird species have been recorded.

Barun valley
Proposed extension to Sagarmatha National Park

Location: Lies directly to the east of the park and in the Arun valley watershed

Biogeographical province: Himalayan Highlands

Physiographic zone: High Himal

Vegetation: Subtropical forests: *Castanopsis tribuloides.* Lower temperate forests: mixed broadleaves, *Quercus lamellosa.* Upper temperate forests: mixed broadleaves, *Rhododendron* spp. Subalpine forests: *Abies spectabilis, Betula utilis, Rhododendron* spp., *Rhododendron* spp./*Juniperus* spp. Alpine zone: *Rhododendron* spp./*Juniperus* spp.

Importance for birds (under-recorded): A species list is given on p.100. Forests in the Barun valley must be among the most important for birds in Nepal. Breeding species recorded total 159 and many more are likely to be found. As many as 66 of these may have significant breeding populations in Nepal. The valley is the only locality in the country for five species: Dark-sided Thrush *Zoothera marginata*, Slaty-bellied Tesia *Tesia olivea*, Broad-billed Warbler *Abroscopus hodgsoni*, Spotted Wren-Babbler *Spelaeornis formosus* and Coral-billed Scimitar-Babbler *Pomatorhinus ferruginosus* and will be the only protected area where Tailed Wren-Babbler *Spelaeornis caudatus* and Rusty-fronted Barwing *Actinodura egertoni* occur. Other notable species are Red-necked Falcon *Falco chicquera*, Satyr Tragopan *Tragopan satyra*, Wood Snipe *Gallinago nemoricola*, Forest Eagle Owl *Bubo nipalensis*, Brown Wood Owl *Strix leptogrammica*,

Rufous-breasted Bush-Robin *Tarsiger hyperythrus*, Long-billed Thrush *Zoothera monticola*, Smoky Warbler *Phylloscopus fuligiventer*, Hill Blue Flycatcher *Cyornis banyumas*, Slaty-backed Flycatcher *Ficedula hodgsonii*, White-gorgetted Flycatcher *F. monileger*, Slender-billed Scimitar-Babbler *Xiphirhynchus superciliaris*, Golden Babbler *Stachyris chrysaea*, Blue-winged Laughing-thrush *Garrulax squamatus*, Scaly Laughing-thrush *G. subunicolor*, Fire-tailed Myzornis *Myzornis pyrrhoura*, Black-headed Shrike-Babbler *Pteruthius rufiventer* and Yellow-bellied Flowerpecker *Dicaeum melanoxanthum*.

Rara Lake National Park

Date gazetted: 1976

Area: 106 sq km

Location: Situated in Mugu district in north-west Nepal

Biogeographical province: Himalayan Highlands

Physiographic zone: High Mountains

Vegetation: Upper temperate forests: *Pinus wallichiana* (sometimes mixed with *Quercus semecarpifolia*), *Juniperus* spp. Subalpine forests: *Abies spectabilis* (sometimes mixed with *Quercus semecarpifolia*), *Betula utilis*, *Juniperus* spp., *Rhododendron* spp. (narrow belt of scrub).

Importance for birds (under-recorded): A species list is given on p.104. The lake, which lies at 3,050 m, is an important staging point for migratory water birds, with 35 species recorded so far, although only small numbers of birds are involved. Breeding species include the western specialities: Cheer Pheasant *Catreus wallichii*, Himalayan Pied Woodpecker *Dendrocopos himalayensis*, White-throated Tit *Aegithalos niveogularis*, Spot-winged Black Tit *Parus melanolophus*, White-cheeked Nuthatch *Sitta leucopsis* and Kashmir Nuthatch *S. cashmirensis*. The park supports 32 breeding species which may have significant populations in Nepal.

Shey-Phoksundo National Park

Date gazetted: 1984

Area: 3,555 sq km

Location: Situated in Dolpo, Jumla and Mugu districts in north-west Nepal, mainly in the trans-Himalayan region

Biogeographical province: Himalayan Highlands

Physiographic zone: High Himal

Vegetation: Upper temperate forests: *Pinus wallichiana*, *Quercus semecarpifolia*. Subalpine forests: *Betula utilis*, *Caragana* spp. Alpine zone: *Caragana* spp.

Importance for birds (under-recorded): A species list is given on p.107. Shey-Phoksundo is important for species typical of trans-Himalayan Nepal, such as Tibetan Partridge *Perdix hodgsoniae*, Brown Accentor *Prunella fulvescens*, Hume's Ground Jay *Pseudopodoces humilis* and Crimson-eared Rosefinch *Carpodacus rubicilloides*. Western specialities include White-throated Tit *Aegithalos niveogularis*, Spot-winged Black Tit *Parus melanolophus*, White-cheeked Nuthatch *Sitta leucopsis* and Kashmir Nuthatch *S. cashmirensis*. A total of 18 breeding species for which Nepal may hold significant populations has been found.

Khaptad National Park

Date gazetted: 1984

Area: 225 sq km

Location: Lies on a plateau at about 3,000 m elevation in far west Nepal

Biogeographical province: Himalayan Highlands

Physiographic zones: Middle Mountains, High Mountains

Vegetation: Subtropical forests: *Pinus roxburghii*, broadleaves. Lower temperate forests: *Quercus lanata/Q. leucotrichophora.* Upper temperate forests: *Quercus semecarpifolia* and mixed *Quercus semecarpifolia/Q. floribunda/Tsuga dumosa/ Abies pindrow.* Subalpine forests: *Abies spectabilis, Rhododendron* spp. Meadows (interspersed with forests on the plateau).

Importance for birds (under-recorded): A species list is given on p.110. A total of 50 species which may have significant breeding populations in Nepal has been recorded up to now, including the Satyr Tragopan *Tragopan satyra* and Great Parrotbill *Conostoma aemodium.* The park is the only protected area in Nepal where the Black-chinned Yuhina *Yuhina nigrimenta* is known to occur. Other notable breeding species include the threatened Brown Wood Owl *Strix leptogrammica*, Yellow-bellied Bush Warbler *Cettia acanthizoides*, Black-throated Parrotbill *Paradoxornis nipalensis* and Fire-capped Tit *Cephalopyrus flammiceps*, and the western specialists Himalayan Pied Woodpecker *Dendrocopos himalayensis* and Spot-winged Black Tit *Parus melanolophus.*

Royal Sukla Phanta Wildlife Reserve

Date gazetted: 1976

Area: 155 sq km

Location: Lies in the lowlands of extreme south-west Nepal

Biogeographical province: Indus-Ganges Monsoon Forest

Physiographic zone: Terai

Vegetation: Tropical forests: *Shorea robusta, Acacia catechu-Dalbergia sissoo.* Grasslands (extensive)

Importance for birds (under-recorded): A species list in given on p.115. Sukla Phanta is important for grassland birds, especially the threatened Swamp Francolin *Francolinus gularis*, Bengal Florican *Houbaropsis bengalensis*, Grass Owl *Tyto capensis*, Large Grass Warbler *Graminicola bengalensis* and Striated Marsh Warbler *Megalurus palustris.* The reserve supports the largest population of Bengal Floricans in Nepal (Inskipp and Inskipp 1983). It is the only locality in the country where Black Bittern *Dupetor flavicollis* regularly occurs. There are 22 breeding species which are at risk in Nepal including Pallas's Fish Eagle *Haliaeetus leucoryphus*, Lesser Fishing Eagle *Ichthyophaga nana*, Grey-headed Fishing Eagle *I. ichthyaetus*, Changeable Hawk-Eagle *Spizaetus cirrhatus*, Brown Fish Owl *Ketupa zeylonensis*, Oriental Pied Hornbill *Anthracoceros coronatus* and Great Slaty Woodpecker *Mulleripicus pulverulentus.* Several specialities of the western lowlands occur such as Sarus Crane *Grus antigone*, Brown-headed Barbet

Megalaima zeylanica, White-naped Woodpecker *Chrysocolaptes festivus* and Tickell's Blue Flycatcher *Cyornis tickelliae*. There is a small lake, Rani Tal, which is valuable for water birds, especially migrating and wintering wildfowl and waders. The reserve is also important for wintering birds. A total of 80 winter visitors has been recorded so far and many more are likely to be found.

Royal Bardia Wildlife Reserve

Date gazetted: 1976

Area: 968 sq km

Location: Situated in mid-western Nepal, mainly in the bhabar zone and also extending into the Siwalik hills

Biogeographical province: Indus-Ganges Monsoon Forest

Physiographic zones: Terai, Siwaliks

Vegetation: Tropical forests: *Shorea robusta* (about 70 percent of the reserve), *Acacia catechu-Dalbergia sissoo, Terminalia-Anogeissus* deciduous hill forest. Subtropical forests: *Pinus roxburghii*. Grasslands and savanna.

Importance for birds (under-recorded): A species list is given on p.121. There are 22 breeding species which are threatened in Nepal including the Rufous-bellied Eagle *Hieraaetus kienerii*, Changeable Hawk-Eagle *Spizaetus cirrhatus*, Pin-tailed Green Pigeon *Treron apicauda*, Forest Eagle Owl *Bubo nipalensis*, Brown Fish Owl *Ketupa zeylonensis*, Oriental Pied Hornbill *Anthracoceros coronatus*, Great Pied Hornbill *Buceros bicornis*, Great Slaty Woodpecker *Mulleripicus pulverulentus*, Silver-eared Mesia *Leiothrix argentauris* and Crow-billed Drongo *Dicrurus annectans*. The grasslands support a small population of the Bengal Florican *Houbaropsis bengalensis*. Lesser Florican *Sypheotides indica* has been recorded and possibly breeds. Western specialities include Grey Francolin *Francolinus pondicerianus*, Sarus Crane *Grus antigone*, Brown-headed Barbet *Megalaima zeylanica*, White-naped Woodpecker *Chrysocolaptes festivus* and Tickell's Blue Flycatcher *Cyornis tickelliae*. The Karnali river valley, which forms the reserve's western boundary, is a migration pathway for wildfowl, notably for Bar-headed Geese *Anser indicus*. The reserve is also important for winter visitors and although only 63 species have been recorded up to now, many more are likely to occur.

Kosi Tappu Wildlife Reserve

Date gazetted: 1976

Area: 175 sq km

Location: Lies in the Sapta-Kosi river plain in the south-east lowlands

Biogeographical province: Bengalian rainforest

Physiographic province: Terai

Vegetation: Once dominated by *Acacia catechu-Dalbergia sissoo* and mixed deciduous forests, but much of these have been degraded by over-exploitation and siltation. Grasslands (extensive). In 1980 the reserve was extended southwards to the Kosi Barrage thereby including a wetland which is of international importance and by far the most valuable in Nepal. However, there is widespread abuse of the regulations.

Importance for birds (under-recorded): A species list is given on p.127. Kosi Tappu is a valuable wintering area and staging point for migrating birds, especially for large numbers of wildfowl, waders, gulls and terns. Winter visitors and passage migrants recorded so far total 125 species and more are likely to be found. There are 18 breeding species which are at risk in Nepal, although two of these: Changeable Hawk-Eagle *Spizaetus cirrhatus* and Dusky Eagle Owl *Bubo coromandus* have not been recorded since 1976 (Dahmer 1976). Other notable threatened species are Swamp Francolin *Francolinus gularis*, Red-necked Falcon *Falco chicquera*, Bengal Florican *Houbaropsis bengalensis*, Brown Fish Owl *Ketupa zeylonensis* and Striated Marsh Warbler *Megalurus palustris*, and is the only protected area in Nepal where the threatened Watercock *Gallicrex cinerea* and Abbott's Babbler *Trichastoma abbotti* are known to occur.

Shivapuri Watershed and Wildlife Reserve

Date gazetted: 1985

Area: 145 sq km

Location: Lies on the northern side of the Kathmandu valley in central Nepal

Biogeographical province: Himalayan Highlands

Physiographic zone: Middle Mountains

Vegetation: About 50 percent of the watershed is still forested. Subtropical forests: *Pinus roxburghii*. Lower temperate forests: *Quercus lanata* (little remains and that is badly degraded). Upper temperate forests: *Quercus semecarpifolia*.

Importance for birds (under-recorded): A species list is given on p.133. Shivapuri supports 26 breeding species for which Nepal may hold significant populations and is valuable for birds of the *Quercus semecarpifolia* forest, notably Yellow-bellied Bush Warbler *Cettia acanthizoides* and Grey-sided Laughing-thrush *Garrulax caerulatus*. The reserve is also important for wintering birds. A total of 36 winter visitors, including 16 which may have significant breeding populations elsewhere in Nepal, has been recorded since 1975 and more are likely to occur. As part of the forest has recently been badly degraded, species recorded before 1975 are listed separately. These include the threatened Forest Eagle Owl *Bubo nipalensis*, White-gorgetted Flycatcher *Ficedula monileger*, Slender-billed Scimitar-Babbler *Xiphirhynchus superciliaris*, Blue-winged Laughing-thrush *Garrulax squamatus* and Cutia *Cutia nipalensis*. Some of these birds could still occur, especially on the northern slopes.

Dhorpatan Hunting Reserve

Date gazetted: 1987

Area: 1,325 sq km

Location: Lies in Baglung district in west central Nepal

Biogeographical province: Himalayan Highlands

Physiographic zone: High Mountains

Vegetation: Lower temperate forests: *Quercus lanata, Pinus wallichiana*. Upper temperate forests: *Pinus wallichiana, Quercus semecarpifolia*. Subalpine forests: *Abies spectabilis, Betula utilis, Juniperus* spp., *Rhododendron* spp./*Juniperus* spp.

Importance for birds (under-recorded): A species list is given on p.137. Dhorpatan has a few western specialities including the Cheer Pheasant *Catreus wallichii* and Himalayan Pied Woodpecker *Dendrocopos himalayensis*. It is the only known locality in Nepal with a good population of the first named species. A total of 41 breeding species for which Nepal may hold significant populations has been recorded so far, including the Satyr Tragopan *Tragopan satyra*.

Annapurna Conservation Area

Proposed protected area, functioning but not officially designated: 2,660 sq km

Location: Lies north of Pokhara in central Nepal

Biogeographical province: Himalayan Highlands

Physiographic zones: High Mountains, High Himal

Vegetation: Subtropical forests: *Schima wallichii/Castanopsis indica*, *Alnus nepalensis*, *Pinus roxburghii*. Lower temperate forests: mixed broadleaves, *Quercus lamellosa*, *Q. lanata*. Upper temperate forests: mixed broadleaves, *Quercus semecarpifolia* (sometimes with *Tsuga dumosa*), *Pinus wallichiana* (pure and mixed with either broadleaves, *Picea smithiana*, *Abies spectabilis* or *Betula utilis*), *Caragana* spp. Subalpine forests: *Abies spectabilis*, *Betula utilis*, *Juniperus* spp., *Caragana* spp. Alpine zone: *Rhododendron* spp./*Juniperus* spp., *Caragana* spp.

Importance for birds (well-recorded): A species list is given on p.140. When it is gazetted the Annapurna Conservation Area will be one of the three most internationally important protected areas for birds in Nepal. Breeding birds total 329, 100 of which may have significant populations in the country. The area also supports 38 breeding species at risk in Nepal. The wide species diversity can be attributed to the large number of habitat types within the area and to its location. The Kali Gandaki which runs through the area is a major divide for bird distribution (see p.41) and species typical both of east and west Nepal occur. The area is of special interest for pheasants as it is the only place where all six Himalayan pheasants occurring in Nepal have been found. It will also be the only protected area where the rare Rufous-throated Partridge *Arborophila rufogularis*, Chestnut-crowned Bush Warbler *Cettia major*, Grey-cheeked Warbler *Seicercus poliogenys*, Pygmy Blue Flycatcher *Muscicapella hodgsoni*, Brown Parrotbill *Paradoxornis unicolor*, Cutia *Cutia nipalensis*, Golden-breasted Fulvetta *Alcippe chrysotis* and Red-browed Finch *Callacanthis burtoni* have been recorded in the breeding season. Other notable breeding species are Wood Snipe *Gallinago nemoricola*, Red-headed Trogon *Harpactes erythrocephalus*, Orange-rumped Honeyguide *Indicator xanthonotus*, Bay Woodpecker *Blythipicus pyrrhotis*, Grey-chinned Minivet *Pericrocotus solaris*, Gould's Shortwing *Brachypteryx stellata*, Rufous-breasted Bush-Robin *Tarsiger hyperythrus*, Long-billed Thrush *Zoothera monticola*, Yellow-bellied Bush-Warbler *Cettia acanthizoides*, Smoky Warbler *Phylloscopus fuligiventer*, Large Niltava *Niltava grandis*, Hill Blue Flycatcher *Cyornis banyumas*, White-gorgetted Flycatcher *Ficedula monileger*, Slender-billed Scimitar-Babbler *Xiphirhynchus superciliaris*, Golden Babbler *Stachyris chrysaea*, Great Parrotbill *Conostoma aemodium*, Fulvous Parrotbill *Paradoxornis fulvifrons*, Rufous-chinned Laughing-thrush *Garrulax rufogularis*, Grey-sided Laughing-thrush *G. caerulatus*, Blue-winged Laughing-thrush *G. squamatus*, Scaly Laughing-thrush *G. subunicolor*, Fire-tailed Myzornis *Myzornis pyrrhoura*, Black-headed Shrike-Babbler *Pteruthius rufiventer*, Yellow-bellied Flowerpecker *Dicaeum melanoxanthum*, Vinaceous Rosefinch *Carpodacus vinaceus*, Crimson-browed Finch *Propyrrhula subhimachala*, Scarlet Finch *Haematospiza sipahi* and Spot-winged Grosbeak *Mycerobas melanozanthos*. The area is valuable for wintering birds too, both resident species and winter visitors which include Saker *Falco cherrug*, Robin Accentor *Prunella rubeculoides*, Güldenstadt's Redstart *Phoenicurus erythrogaster*, Chestnut Thrush *Turdus rubrocanus* and Pine Bunting *Emberiza leucocephalos*. A total of 50 winter visitors has been recorded, including six which may have significant breeding populations in other parts of Nepal.

The Kali Gandaki is an important migration pathway in autumn. Thousands of cranes, mainly Demoiselle Cranes *Anthropoides virgo* and about 40 other bird species, including nearly 20 raptors, have been recorded. Larger numbers of birds of prey have recently been seen totalling over 8,000 individuals of about 20 species migrating west through the area. The birds followed the southern ridges of the Himalayan range (de Roder in press).

AFFORESTATION

The problems of forest losses and deterioration have been recognised for a long time. Up to 1956 many local communities had rules for forest protection and management, although these rules had no legal basis other than custom (Jackson 1987). In 1957 His Majesty's Government of Nepal nationalised Nepal's forests in an attempt to protect them. Unfortunately the people believed that nationalisation meant the government had taken away their forests and that continuation of their old systems of management was futile. Much uncontrolled forest exploitation resulted.

Recognising that forest management must ultimately rest with local communities, the government introduced the National Forestry Plan (NAFP 1979) in 1976. This provides for increased afforestation and improved protection and management of natural forests. In 1978 new regulations (see p.9) enabled forests to be handed back to the panchayats (councils representing groups of villages with an average population of about 2,000). Despite this legislation it is proving difficult to return control of forests to local people. This is partly because they have not been able to use their own initiative for about 30 years, but there is also a lack of trained foresters who are sympathetic towards community forestry. Nevertheless, there has recently been a great expansion of forestry activities in the panchayats (Jackson 1987). Carter (1987) lists 24 afforestation organisations, including six major ones: Community Forestry and Afforestation Division, Community Forestry Development Project, Hill Forest Development Project, Kosi Hills Rural Development, Nepal-Australia Forestry Project and Terai Community Forestry Project. By mid-1986, 714 sq km of land had been planted or managed as protected forest (Carter 1987), but a much greater area is required eventually. The World Bank (1978) has estimated that in the next 20 years 6,400 sq km of village woodlots should be planted in the hills and 4,000 sq km in the terai. An additional 2,930 sq km will be required by the larger towns and industry, but the community forestry programme has not been designed to cope with this extra demand (Jackson 1987).

Most of the plantations are of pines mainly *Pinus roxburghii* and *P. wallichiana* which grow throughout much of Nepal, and some are of the exotic *P. patula*. Pines were used because the plantation land was so degraded that these were the only trees that would grow. The successful planting of pines, carried out by the Nepal-Australia Forestry Project (NAFP) has shown that land can be reforested and has gained a great deal of local trust and support (Sattaur 1987). However, pines are much less valuable to local communities than are broadleaved trees because conifer foliage is not suitable for animal fodder or bedding and the wood is not as popular for fuel as the wood from most broadleaves (Shakya and Thompson 1987).

As far as most wildlife is concerned plantations of native pines must be preferable to areas devoid of forest. The areas available for planting are relatively small as the regulations enable a maximum of 1.3 sq km of bare or sparsely forested government land to be allocated to each panchayat. Planting up these areas would both improve and diversify the habitat. However, bird communities of pine forests are species-poor and those in indigenous broadleaves are much more diverse.

In time the plantations could well contain more native broadleaved trees. Indigenous broadleaves, such as *Schima wallichii* have sprung up among the early NAFP pine plantations. Research carried out by the NAFP has shown that if pines which are competing with broadleaves are gradually removed, a pine plantation can eventually be converted to one of broadleaves. This may take as long as 20 years in some places, but a much shorter time in others.

However, new plantations, even of native broadleaves, can never replace the richness and variety of the natural forests which may have taken hundreds of years to develop. A natural forest provides innumerable microhabitats for wildlife, such as trees and shrubs of different species and ages, glades and decaying logs and stumps. A wide range of conditions are available, such as sunlight, deep shade and humidity for example. In contrast the plantations are less valuable for wildlife because their structure and tree species composition are relatively uniform. Trees of the same age and species are frequently planted in straight lines. Up to now most effort has been put into tree planting, but protection of some severely degraded natural forests has resulted in dramatic recoveries and forests well-stocked with valuable native species (Jackson 1987).

Government regulations allow up to 5.2 sq km of natural forest to be protected and managed to meet the needs of the local population in a panchayat. This figure is four times the area which can be allocated for plantations. Unfortunately, there is a psychological barrier to be overcome: local people tend to regard natural forest as land which is not being used; they will more readily protect an area in which trees have been planted (Jackson 1987). It has been found that low density enrichment planting can enhance protection because villagers recognise that their forests have been improved and subsequently exercise more restraint in their use of the forest.

The Forestry Research and Information Centre in Nepal believe that improved management of the large existing areas of low density forest (control of cutting, lopping, grazing, etc.) is the most important aspect of Nepalese forestry (Anon. 1987). This policy has enormous potential and would also be much more beneficial to forest birds than planting more trees.

Slender-billed Scimitar-Babbler *Xiphirhynchus superciliaris* (Linda Schrijver)

Wood Snipe *Gallinago nemoricola* (Dave Showler)

THREATS

FOREST LOSSES AND DETERIORATION

Nepal's forest resources can no longer meet the vital requirements of its people for fuel, animal fodder and other basic materials. As the population is estimated to be rising at 2.67 percent per year (HMG/IUCN 1983) pressure on the forests is increasing. The resulting losses and deterioration of forests are the major threats facing wildlife in Nepal.

Whilst the people's needs have caused most deforestation, tourist trekkers and mountaineering expeditions are exacerbating the problem in some places by often unnecessary demands for wood. Surveys indicate that the firewood used by a tourist trekker is nearly five times more than is required by a local Sherpa in Khumbu (Mishra 1986). Mountaineering expeditions use even more wood than trekkers. It has been estimated that three major mountaineering expeditions in one year consumed 25,000 to 30,000 kg of firewood (Mishra 1986). Some of the forests most affected lie within Nepal's national parks, notably Sagarmatha (Mount Everest) National Park which is also a World Heritage site.

A few forests which are especially important for birds are badly threatened. The trail between Ghorepani and Ghandrung on the southern flanks of Annapurna is an example. This route was once almost never used by local people and an extensive unbroken oak/rhododendron forest, important for rare species such as the Orange-rumped Honeyguide *Indicator xanthonotus*, covered the surrounding ridges. Within the last eight years the trail has become a popular trekking route with half a dozen large forest clearings created for tourist lodges by 1986 (Inskipp 1987). Forests from the terai up to the limit of permanent large scale cultivation (about 2,400 m) are under the greatest pressure as they are in demand throughout the year, in contrast to forests higher up which are mainly utilised between May and October.

Forest losses

Valuable data on Nepal's forests are provided by the Land Reserves Mapping Project (LRMP), largely based on 1:50,000 air photographs taken in 1978/79. In addition false colour satellite imagery at a scale of 1:250,000 was used. This gave useful information for the High Himal which was not covered by conventional air photography.

Once Nepal was extensively forested, but by 1979 only 42.8 percent of the country was 'forest land' (partially covered in trees and shrubs). A large percentage of this, however, was covered in forest of low density. The measurements show that only 28.1 percent of the country had more than 40 percent tree cover, 10 percent had only 10-40 percent tree cover and 4.7 percent had only shrub cover (Kenting 1986). Forest losses between 1965 and 1979 are given in Table 4, based on air photographs (Kenting 1986). The figure of 5.7

Table 4: Comparison of estimated forest areas in 1964/65 and 1978/79.

Physiographic region	WECS (1964/65 photos)	LRMP (1978/79 photos)	Change from 1964/65 sq km	percent
	sq km	sq km		
High Himal	2,218*	2,218	-	-
High and Middle Mountains	39,438	40,162	+724	+1.8
Siwaliks	17,393	14,760	-2,633	-15.1
Terai	7,838	5,929	-1,911	-24.4
Total	66,887	63,069	3,820	-5.7

LRMP = Land Resources Mapping Project   * LRMP figure
WECS = Water and Energy Commission Survey   Source: Kenting (1986)

percent loss of forest for all the country during the period is considered to be reasonably accurate, although on the conservative side (Kenting 1986).

Nearly all of these losses took place in the terai and Siwaliks (Table 4) (Kenting 1986). Once these regions were sparsely inhabited because of the high risk of malaria. The dramatic reduction of the disease and the high population in the Middle Mountains resulted in mass migration of people to the terai and dun valleys. The Nepal Government is organising resettlement schemes, but squatters are by far the largest group of settlers.

Forests in the terai and Siwaliks have mainly been felled to make way for agriculture. It was suggested by Rieger (1981) that if migration into the terai continued at the rate of the previous ten years all the good farmland would be occupied in little more than a decade. Further clearance for cultivation in the Siwalik zone is also likely to be reduced in the future because the cultivated land lies in the dun valleys and these are of limited extent.

Throughout the terai forests are also being felled by commercial logging, but the areas involved have been small compared to those cleared for agriculture. However, tropical evergreen forest, an unprotected forest type of very small extent in Nepal was being severely depleted by logging in 1988 (van Riessen pers. comm.). Details of logging operations are given in Carter (1987).

There are three carrying out logging operations companies in Nepal. One of the largest, the Sagarnath Forest Development Project felled 50 sq km of forest during the first phase of operations between 1979 and 1986. This comprised only 0.26 percent of the terai forest losses measured between 1965 and 1979 by the LRMP. In addition, an unknown quantity of timber is being illicitly felled in terai forests and smuggled across the border to India (Oy and Madecor 1987b). The clear-felled forests have been partly replaced by plantations which total over 70 sq km, although at least 21.5 sq km are of eucalypts *Eucalyptus* spp.

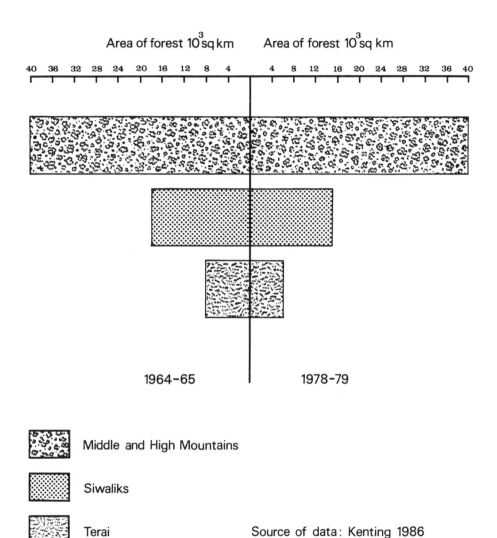

Figure 4: Comparison of estimated forest areas in 1964/65 and 1978/79.

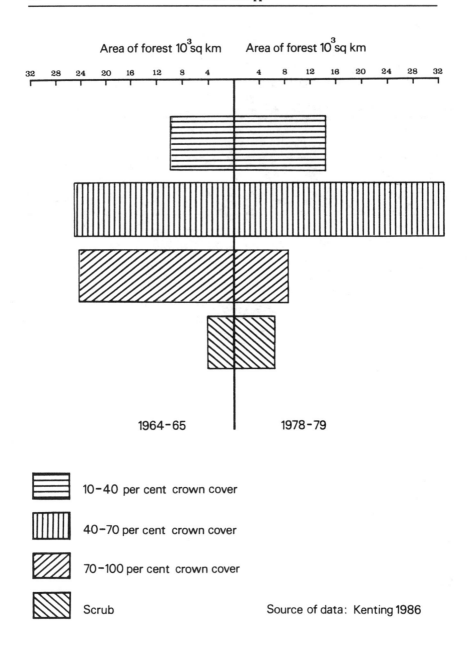

Figure 5: Estimates of changes in forest tree crown cover between 1964/65 and 1978/79.

The gain of 724 sq km for the Middle and High Mountains between 1965 and 1979 (Table 4) is not considered statistically significant because the earlier analysis was based on estimates. Contrary to popular belief deforestation in the Middle Mountains is not a recent phenomenon and conversion of forested to cultivated land was well established by at least the late eighteenth century (Mahat et al. 1986).

Figures for the High Himal were not available in 1965, but there was undoubtedly some loss of forest land between 1965 and 1979 (Kenting 1986).

Oy and Madecor (1987b) estimated that total forest and scrub cover throughout Nepal declined by a further 0.6 percent between 1979 and 1986. Almost all the losses were thought to have occurred in the terai.

Forest deterioration

The LRMP has used crown density as a measure of forest stocking. It is expressed as a percentage of ground area covered by tree crowns when viewed from the air or from an aerial photograph. Forest with 10-40 percent crown cover is considered poorly stocked. The deterioration of Nepal's forests between 1965 and 1979 can be clearly seen from Table 5 and Figure 5. The conditions of almost all Nepal's forests are considered to be worsening. The exceptions are forests in very remote areas, such as the Barun valley (Shrestha et al. 1985) and subalpine forests of the Karnali region in north-west Nepal (Shrestha 1984-1985). The most affected region is the Middle Mountains of the centre and east. Although the area of forested land in the region did not significantly alter during the period, forest quality deteriorated more than anywhere else in the country. As much as 60 percent of the region's forested land had only 10-40 percent tree cover in 1979.

Table 5: Estimates of changes in forest tree crown cover between 1964/65 and 1978/79.

Category	Crown cover of forest land			
	10-40%	40-70%	70-100%	Shrub
Forest land area 1964/65 (sq km)	10,218	25,544	25,027	3,888
Forest land area 1978/79 (sq km)	14,187	32,178	8,251	6,233
Changes between 1964/65 and 1978/79 (sq km)	+3,969	+6,634	-16,776	+2,350

Source: Kenting (1986)

Forests are depleted in the following ways. Tree branches are cut for fuel and foliage; oaks *Quercus* spp. are often heavily lopped to provide animal fodder and bedding, frequently reducing forests to low coppice 2-3m in height. Lopping often leads to the stunting and finally the death of trees and an impoverishment of the variety of species as only some trees can survive this treatment (Rieger 1981). Large quantities of bamboo *Arundinaria* spp. and *Bambusa* spp. are cut for weaving mats and baskets and for construction work. Excessive numbers of livestock are held in Nepal and forests are used for grazing throughout the year. Livestock feed on young plants, leaves and twigs of small trees, often severely reducing the understorey and even preventing forest regeneration altogether. In many forests the undergrowth and ground layer are burned regularly, often annually, to improve the growth of grasses. This practice favours the spread of fire-resistant species, such as pines. Pines are often succeeded by broadleaved trees, but frequent fires prevent this. The result is an open forest of old pines lacking undergrowth and only supporting a low variety of bird species. The understorey in many forest types is suffering from uncontrolled cutting, over-grazing or burning. Its loss or reduction must drastically affect the bird species composition. Many species, including babblers, warblers, chats and thrushes inhabit this part of the forest ecosystem. Unfortunately it has not been possible to assess the condition of the forest understorey from air photographs. Some tree species, such as oaks, are favoured for construction work and so are felled in preference to others. Selective felling over a long period can change the forest composition and also its wildlife. During a study of a seemingly healthy forest in the Arun valley, Cronin was surprised at the lack of certain bird species. He found that local villagers had felled oak in preference to other species. As a result *Castanopsis* trees had become dominant and a drier forest had been produced with a lower variety of plants and birds than previously (Cronin 1979).

Besides the opening up of forests by selective felling, removal of foliage and lopping of branches must also result in their becoming unsuitable for the numerous species which require dense or moist forests.

Species which prefer open forests, such as some flycatchers, may well have benefited as a result of forest thinning. Presumably birds which frequent scrub, such as the White-cheeked Bulbul *Pycnonotus leucogenys*, Grey Bushchat *Saxicola ferrea* and Striated Prinia *Prinia criniger* must have also increased in recent years. However, most species in this category are common and widespread in Nepal, whilst many forest birds are declining. Overall forest depletion can have benefited relatively few species (see p.43) and the populations of most Nepalese forest birds are likely to have decreased.

Effects of deforestation: soil erosion and flooding

Natural erosion rates in Nepal are very high, especially during the monsoon rains. Over the last century, however, an increasing proportion of soil loss is attributable to the increased population pressure on a limited land resource by deforestation and over-grazing (Carson 1985). Ultimately mountain slopes become both uncul-tivable and no longer able to support trees. Rapid run-off of rain during the monsoon has also led to widespread flooding in the lowlands. One disaster took place in September 1981 when a landslide temporarily blocked the Tinnau river north of Butwal during heavy prolonged rainfall. The subsequent breakthrough of this dam resulted in a downstream surge that killed almost two hundred people (Carson 1985).

PESTICIDES

In the 1950s large quantities of DDT and dieldrin were sprayed in the lowlands in an attempt to eradicate malaria. Although no studies have been made, it is likely that bird populations suffered, especially those of birds of prey, as a result of the chemicals entering the food chain.

HUNTING AND PERSECUTION

The effects of hunting and persecution on bird populations are unknown, but are probably much less than habitat losses. Local hunting pressures can be very high, such as in the Himalaya south of the Annapurna Himal (Lelliott and Yonzon 1979). Here amateur and professional hunters trap and shoot pheasants for food throughout the year. Flashlight shooting is particularly effective and can wipe out all the Kalij Pheasants *Lophura leucomelana* in an area over a short time period. A survey of trapping carried out in 1979 revealed that tolls comprised Kalij Pheasants 43 percent, Satyr Tragopans *Tragopan satyra* 36 percent and Himalayan Monals *Lophophorus impejanus* 21 percent (Lelliott and Yonzon 1979). Six pheasants occur at Ghasa in the Kali Gandaki valley. Cheer Pheasant *Catreus wallichii* suffers more than the others, probably because it roosts communally and closer to human habitation (S. Gawn *in litt.* 1987). Roberts (1979) discussed the effects of hunting throughout the country and pointed out that human interference in sparsely populated west Nepal is likely to be low; in the highlands east of Kathmandu the people are mainly Buddhist and those who adhere strongly to the faith have an objection to taking life.

Nepalese quails may have been adversely affected by hunting. Almost all of them were more common last century (Inskipp and Inskipp 1985). Hodgson, for instance, obtained at least six specimens of the Black-breasted Quail *Coturnix coromandelicus* from the Kathmandu valley (Hodgson 1829,1837), but there are only two later records for the country, both in the 1950s (Fleming and Traylor 1961; Krabbe 1983).

Raptors may be persecuted near villages as they sometimes take chickens. In 1986 the skins of a Mountain Hawk-Eagle *Spizaetus nipalensis* and a Crested Serpent Eagle *Spilornis cheela*, apparently shot for this reason, were found at a lodge at Pothana, north-west of Pokhara (pers. obs.)

Spot-winged Grosbeak *Mycerobas melanozanthos* (Dave Showler)

NEPALESE FOREST BIRDS

OVERVIEW

The status and distribution of Nepal's birds were described in Inskipp and Inskipp (1985) and species which may have significant breeding populations in the country (all habitats) were listed in Inskipp and Inskipp (1986). Since then an additional 17 species have been recorded in the country including three breeding forest species: Slaty-bellied Tesia *Tesia olivea*, Broad-billed Warbler *Abroscopus hodgsoni* and Spotted Wren-Babbler *Spelaeornis formosus*, which were all located in the upper Arun watershed by H. S. Nepali (Shrestha *et al.* 1985). Nepal may be especially important for the latter two species as they are rare and have particularly restricted ranges in the Indian subcontinent. Breeding species for which Nepal may hold internationally significant populations and which are confined to forests are listed in Table 6.

The 469 species comprising the breeding bird communities in Nepal's main forest types and symbols denoting their updated status are listed in the Appendix. Data were taken from the references listed on p.161. Analyses of these breeding bird communities and the most important forests for breeding birds are described on p.43.

Table 6: Breeding forest birds for which Nepal may hold internationally significant populations.

    Blood Pheasant *Ithaginis cruentus*
    Satyr Tragopan *Tragopan satyra*
*   Cheer Pheasant *Catreus wallichii*
*   Wood Snipe *Gallinago nemoricola*
    Speckled Woodpigeon *Columba hodgsonii*
    Ashy Woodpigeon *Columba pulchricollis*
    Slaty-headed Parakeet *Psittacula himalayana*
*   Orange-rumped Honeyguide *Indicator xanthonotus*
    Himalayan Pied Woodpecker *Dendrocopos himalayensis*
    Darjeeling Pied Woodpecker *Dendrocopos darjellensis*
    Crimson-breasted Pied Woodpecker *Dendrocopos cathpharius*
    Brown-fronted Pied Woodpecker *Dendrocopos auriceps*
*   Blue-naped Pitta *Pitta nipalensis*
    Striated Bulbul *Pycnonotus striatus*
    Rufous-breasted Accentor *Prunella strophiata*
    Robin Accentor *Prunella rubeculoides*
*   Gould's Shortwing *Brachypteryx stellata*
    Indian Blue Robin *Luscinia brunnea*

Table 6   (contd)

Golden Bush-Robin *Tarsiger chrysaeus*
White-browed Bush-Robin *Tarsiger indicus*
*   Rufous-breasted Bush-Robin *Tarsiger hyperythrus*
Blue-fronted Redstart *Phoenicurus frontalis*
White-throated Redstart *Phoenicurus schisticeps*
White-bellied Redstart *Hodgsonius phoenicuroides*
*   Purple Cochoa *Cochoa purpurea*
Blue-capped Rock-Thrush *Monticola cinclorhyncha*
Plain-backed Mountain Thrush *Zoothera mollissima*
Long-tailed Mountain Thrush *Zoothera dixoni*
*   Long-billed Thrush *Zoothera monticola*
*   Pied Ground Thrush *Zoothera wardii*
Tickell's Thrush *Turdus unicolor*
White-collared Blackbird *Turdus boulboul*
Black-backed Forktail *Enicurus immaculatus*
Chestnut-headed Tesia *Tesia castaneocoronata*
*   Pale-footed Bush Warbler *Cettia pallidipes*
*   Chestnut-crowned Bush Warbler *Cettia major*
Aberrant Bush Warbler *Cettia flavolivacea*
Grey-sided Bush Warbler *Cettia brunnifrons*
*   Grey-capped Prinia *Prinia cinereocapilla*
Grey-cheeked Warbler *Seicercus poliogenys*
Grey-hooded Warbler *Seicercus xanthoschistos*
*   Broad-billed Warbler *Abroscopus hodgsoni*
Black-faced Warbler *Abroscopus schisticeps*
Large-billed Leaf Warbler *Phylloscopus magnirostris*
Orange-barred Leaf Warbler *Phylloscopus pulcher*
*   Smoky Warbler *Phylloscopus fuligiventer*
Rufous-bellied Niltava *Niltava sundara*
Rufous-tailed Flycatcher *Muscicapa ruficauda*
Sapphire Flycatcher *Ficedula sapphira*
Ultramarine Flycatcher *Ficedula superciliaris*
Slaty-backed Flycatcher *Ficedula hodgsonii*
Yellow-bellied Fantail *Rhipidura hypoxantha*
Slender-billed Scimitar-Babbler *Xiphirhynchus superciliaris*
Greater Scaly-breasted Wren-Babbler *Pnoepyga albiventer*
*   Spotted Wren-Babbler *Spelaeornis formosus*
*   Tailed Wren-Babbler *Spelaeornis caudatus*
Black-chinned Babbler *Stachyris pyrrhops*
*   Great Parrotbill *Conostoma aemodium*
*   Brown Parrotbill *Paradoxornis unicolor*
*   Fulvous Parrotbill *Paradoxornis fulvifrons*
*   Spiny Babbler *Turdoides nipalensis*
White-throated Laughing-thrush *Garrulax albogularis*
Striated Laughing-thrush *Garrulax striatus*
Variegated Laughing-thrush *Garrulax variegatus*
Rufous-chinned Laughing-thrush *Garrulax rufogularis*
Spotted Laughing-thrush *Garrulax ocellatus*
Grey-sided Laughing-thrush *Garrulax caerulatus*

Table 6 (contd)

Rufous-necked Laughing-thrush *Garrulax ruficollis*
Blue-winged Laughing-thrush *Garrulax squamatus*
Scaly Laughing-thrush *Garrulax subunicolor*
Black-faced Laughing-thrush *Garrulax affinis*
* Fire-tailed Myzornis *Myzornis pyrrhoura*
* Black-headed Shrike-Babbler *Pteruthius rufiventer*
Green Shrike-Babbler *Pteruthius xanthochloris*
Rusty-fronted Barwing *Actinodura egertoni*
* Hoary Barwing *Actinodura nipalensis*
Red-tailed Minla *Minla ignotincta*
* Golden-breasted Fulvetta *Alcippa chrysotis*
White-browed Fulvetta *Alcippe vinipectus*
Nepal Fulvetta *Alcippe nipalensis*
Black-capped Sibia *Heterophasia capistrata*
Whiskered Yuhina *Yuhina flavicollis*
Stripe-throated Yuhina *Yuhina gularis*
Rufous-vented Yuhina *Yuhina occipitalis*
Black-browed Tit *Aegithalos iouschistos*
* White-throated Tit *Aegithalos niveogularis*
Grey-crested Tit *Parus dichrous*
Rufous-vented Black Tit *Parus rubidiventris*
Spot-winged Black Tit *Parus melanolophus*
White-cheeked Nuthatch *Sitta leucopsis*
White-tailed Nuthatch *Sitta himalayana*
Kashmir Nuthatch *Sitta cashmirensis*
* Rusty-flanked Treecreeper *Certhia nipalensis*
Fire-capped Tit *Cephalopyrus flammiceps*
Fire-tailed Sunbird *Aethopyga ignicauda*
* Yellow-bellied Flowerpecker *Dicaeum melanoxanthum*
Grey-backed Shrike *Lanius tephronotus*
Lanceolated Jay *Garrulus lanceolatus*
Yellow-billed Blue Magpie *Urocissa flavirostris*
* Spot-winged Starling *Saroglossa spiloptera*
Red-browed Finch *Callacanthis burtoni*
* Crimson Rosefinch *Carpodacus rubescens*
Dark-breasted Rosefinch *Carpodacus nipalensis*
Pink-browed Rosefinch *Carpodacus rhodochrous*
* Vinaceous Rosefinch *Carpodacus vinaceus*
Dark-rumped Rosefinch *Carpodacus edwardsii*
* Spot-winged Rosefinch *Carpodacus rhodopeplus*
White-browed Rosefinch *Carpodacus thura*
Crimson-eared Rosefinch *Carpodacus rubicilloides*
* Crimson-browed Finch *Propyrrhula subhimachala*
* Scarlet Finch *Haematospiza sipahi*
Gold-naped Finch *Pyrrhoplectes epauletta*
Red-headed Bullfinch *Pyrrhula erythrocephala*
Collared Grosbeak *Mycerobas affinis*
* Spot-winged Grosbeak *Mycerobas melanozanthos*

* indicates species for which Nepal may be especially important

It must be noted that a considerable proportion (36 percent) of Nepal's breeding species are altitudinal migrants which spend much of their time outside the breeding season in forests lower down and usually in a different climatic zone. In addition there are 33 winter visitors to Nepal's forests (Table 7). To effectively conserve all Nepal's forest birds it is therefore important to also protect forests utilised in the non-breeding season. The identification of forests important for wintering birds requires a separate study. Determining birds' preferred habitats in the non-breeding season is more difficult than identifying breeding habitats. For instance adverse weather conditions may force birds to move to altitudes below their normal winter range. Another complication is that records of species outside the non-breeding season may be of altitudinal migrants moving up or down.

Table 7: Winter visitors to Nepal's forests (excluding species which are also resident).

	Abundance
Greater Spotted Eagle *Aquila clanga*	Uncommon
Common Woodpigeon *Columba palumbus*	Rare
Eurasian Wryneck *Jynx torquilla*	Occasional
Maroon-backed Accentor *Prunella immaculata*	Occasional
Black-throated Accentor *Prunella atrogularis*	Fairly Common
Rufous-backed Redstart *Phoenicurus erythronotus*	Occasional
Hodgson's Redstart *Phoenicurus hodgsoni*	Occasional
Eurasian Blackbird *Turdus merula*	Uncommon
Chestnut Thrush *Turdus rubrocanus*	Uncommon
Eye-browed Thrush *Turdus obscurus*	Rare
Rufous-tailed Thrush *Turdus naumanni*	Rare
Dark-throated Thrush *Turdus ruficollis*	Common
Blyth's Reed Warbler *Acrocephalus dumetorum*	Fairly Common
Thick-billed Warbler *Acrocephalus aedon*	Uncommon
Lesser Whitethroat *Sylvia curruca*	Uncommon
Yellow-faced Warbler *Phylloscopus cantator*	Uncommon
Dusky Warbler *Phylloscopus fuscatus*	Occasional
Sulphur-bellied Warbler *Phylloscopus griseolus*	Rare
Chiffchaff *Phylloscopus collybita*	Occasional
Red-breasted Flycatcher *Ficedula parva*	Common
Brown Shrike *Lanius cristatus*	Fairly Common
Bay-backed Shrike *Lanius vittatus*	Uncommon
Common Chaffinch *Fringilla coelebs*	Occasional
Brambling *Fringilla montifringilla*	Uncommon
Tibetan Serin *Serinus thibetanus*	Fairly Common
Black-faced Bunting *Emberiza spodocephala*	Uncommon
Pine Bunting *Emberiza leucocephalos*	Fairly Common
Yellowhammer *Emberiza citrinella*	Rare

EXTINCT SPECIES

A total of 20 forest species which were originally mainly collected by Hodgson are probably extinct in Nepal (Table 8) (Inskipp and Inskipp 1985). None has been recorded since the last century, with the possible exception of the Black-tailed Crake *Porzana bicolor*, which may have been obtained in the extreme east of the country in 1912 (Stevens 1925b). It is likely that seven of them: Black-backed Kingfisher *Ceyx erithacus*, Blue-fronted Robin *Cinclidium frontale*, White-spectacled Warbler *Seicercus affinis*, Long-billed Wren-Babbler *Rimator malacoptilus*, Red-headed Parrotbill *Paradoxornis ruficeps*, Yellow-throated Fulvetta *Alcippe cinerea* and Collared Treepie *Dendrocitta frontalis* were not collected in Nepal, but actually in forests over the eastern border. Hodgson made two collections of birds, the second while he was based at Darjeeling in India. Although birds from the second collection were listed as coming from Nepal in Gray's published catalogue of the collection (1863), Hodgson deleted Nepal as a source of these donations in an annotated copy of the catalogue (Cocker and Inskipp 1988).

Table 8: Extinct Nepalese forest birds.

	Imperial Heron *Ardea imperialis*
	Jungle Bush Quail *Perdicula asiatica*
	Black-tailed Crake *Porzana bicolor*
	Hodgson's Hawk-Cuckoo *Hierococcyx fugax*
	Oriental Bay Owl *Phodilus badius*
*	Black-backed Kingfisher *Ceyx erithacus*
	Rufous-necked Hornbill *Aceros nipalensis*
	Silver-breasted Broadbill *Serilophus lunatus*
*	Blue-fronted Robin *Cinclidium frontale*
	Green Cochoa *Cochoa viridis*
	Rufous Prinia *Prinia rufescens*
	Mountain Tailorbird *Orthotomus cuculatus*
*	White-spectacled Warbler *Seicercus affinis*
*	Long-billed Wren-Babbler *Rimator malacoptilus*
	Black-breasted Parrotbill *Paradoxornis flavirostris*
*	Red-headed Parrotbill *Paradoxornis ruficeps*
	Red-faced Liocichla *Liocichla phoenicea*
*	Yellow-throated Fulvetta *Alcippe cinerea*
	Long-tailed Sibia *Heterophasia picaoides*
*	Collared Treepie *Dendrocitta frontalis*

* indicates species in Hodgson's second collection

There is also some doubt over whether Hodgson obtained Rufous Prinia *Prinia rufescens*, Mountain Tailorbird *Orthotomus cuculatus*, Black-breasted Parrotbill *Paradoxornis flavirostris* and Red-faced Liocichla *Liocichla phoenicea* in Nepal. The first three species are listed for Nepal by Sharpe (1883) and the last by Gould (1837-1838), but none of them is included for Nepal in the catalogues of Hodgson's collections (Gray and Gray 1846; Gray 1863), nor are they mentioned in his published papers or original notes. The record of Oriental Bay Owl *Phodilus badius* is of a specimen bought from a Kathmandu shop (Hodgson 1829) and it is possible that this bird may also not have originated in Nepal.

Hodgson gave very few localities for specimens of the birds now presumed to be extinct in Nepal, so they have been assigned to forest zones in the Appendix according to their present distribution in India (Ali and Ripley 1984). The two parrotbills inhabit dense thickets of grass, reeds and bamboo in the tropical and subtropical zones. Dense, damp evergreen forests are required by all of the other species except Jungle Bush Quail *Perdicula asiatica*, which frequents secondary forest growth and grass and scrub jungle. All of these species, except for the above-mentioned one, reach the western limit of their historical range in Nepal. Suitable habitat once occurred for them from the centre of the country eastwards, but very little now remains. However, it is possible that some could still occur in far eastern Nepal.

BIRD DISTRIBUTION

On its way from the Tibetan plateau to the Indian plains the Kali Gandaki river has gouged the world's deepest river valley right through the Himalayan range. The river runs north/south through almost the middle of Nepal and the centre of the Himalayan chain. In general, forests to the east of the valley are wetter, richer in plant species and have a greater number of plant epiphytes than western forests. Conifers are much more widespread in the west and rhododendrons in the east.

Black-faced Laughing-thrush *Garrulax affinis* (Linda Schrijver)

Table 9: Distributional limits of breeding Nepalese forest birds.

Column 1: Species likely to occur in west Nepal (recorded from Kali Gandaki valley and further east, and also west of Nepal in India)
Column 2: Species recorded from Kali Gandaki valley and further east
Column 3: Species recorded from Kali Gandaki valley and further west
Column 4: Species recorded from Arun valley and further east

o indicates species whose western limit of world range apparently lies in the Kali Gandaki valley or its watershed.

NB Breeding season records have been included

	1	2	3	4
Hieraaetus pennatus	x			
Hieraaetus fasciatus	x			
Falco chicquera	x			
Falco severus	x			
Coturnix coromandelica	x			
Catreus wallichii			x	
Gallinago nemoricola	x			
Columba pulchricollis		o x		
Macropygia unchall	x			
Treron curvirostra		o x		
Ducula badia		x		
Loriculus vernalis		x		
Clamator coromandus	x			
Chrysococcyx maculatus	x			
Bubo coromandus	x			
Ketupa flavipes	x			
Harpactes erythrocephalus	x			
Halcyon coromanda		x		
Alcedo hercules		x		x
Alcedo meninting		x		
Megalaima franklinii		o x		
Megalaima australis		x		x
Indicator xanthonotus	x			
Sasia ochracea	x			
Gecinulus grantia		x		x
Dendrocopos himalayensis			x	
Dendrocopos darjellensis		o x		
Dendrocopos cathpharius		o x		
Pitta nipalensis		x		
Pitta sordida	x			
Pericrocotus brevirostris		o x		
Pericrocotus solaris		o x		
Pycnonotus striatus		o x		
Criniger flaveolus		x		
Irena puella		x		
Brachypteryx stellata	x			
Brachypteryx montana	x			

Table 9 (contd)

	1	2	3	4
Brachypteryx leucophrys	x			
Tarsiger hyperythrus		o x		
Cinclidium leucurum		o x		
Cochoa purpurea	x			
Zoothera monticola	x			
Zoothera marginata		x		
Turdus viscivorus			x	
Tesia olivea		x		x
Cettia major	x			
Bradypterus thoracicus	x			
Prinia atrogularis		x		x
Seicercus poliogenys	x			
Seicercus castaniceps	x			
Abroscopus hodgsoni		x		x
Abroscopus albogularis		x		x
Abroscopus superciliaris		o x		
Phylloscopus occipitalis			x	
Phylloscopus fuligiventer		x		
Niltava grandis		o x		
Cyornis unicolor	x			
Cyornis banyumas		x		
Muscicapella hodgsoni		o x		
Muscicapa ferruginea		o x		
Ficedula sapphira		x		x
Ficedula hodgsonii		o x		x
Ficedula hyperythra	x			
Ficedula monileger		o x		
Pellorneum ruficeps	x			
Trichastoma abbotti		x		x
Pomatorhinus ferruginosus		x		x
Xiphirhynchus superciliaris		o x		
Spelaeornis formosus		x		x
Spelaeornis caudatus		x		x
Stachyris ruficeps		x		x
Stachyris chrysaea		o x		
Stachyris nigriceps		o x		
Paradoxornis unicolor		o x		
Paradoxornis fulvifrons		o x		
Garrulax monileger		o x		
Garrulax pectoralis		o x		
Garrulax caerulatus		o x		
Garrulax ruficollis		x		
Garrulax squamatus		o x		
Garrulax subunicolor		o x		
Myzornis pyrrhoura		o x		
Cutia nipalensis	x			
Pteruthius rufiventer		o x		

Table 9 (contd)

	1	2	3	4
Pteruthius melanotis		o x		
Gampsorhynchus rufulus		x		x
Actinodura egertoni		x		
Minla ignotincta		o x		
Alcippe chrysotis		o x		
Alcippe castaneceps		o x		
Alcippe nipalensis		o x		
Heterophasia annectans		x		x
Yuhina bakeri		x		x
Yuhina occipitalis		o x		
Aegithalos iouschistos		o x		
Aegithalos niveogularis			x	
Parus rufonuchalis			x	
Parus melanolophus			x	
Parus ater		o x		
Parus spilonotus		x		x
Melanochlora sultanea		x		
Sitta cashmirensis			x	
Certhia discolor	x			
Anthreptes singalensis		x		
Arachnothera longirostra		x		
Arachnothera magna		x		
Dicaeum chrysorrheum		x		x
Dicaeum melanoxanthum	x			
Dicaeum concolor		o x		
Dicaeum cruentatum		x		x
Artamus fuscus	x			
Gracula religiosa	x			
Callacanthis burtoni	x			
Carpodacus rubescens		x		
Carpodacus edwardsii		x		
Carpodacus rhodopeplus	x			
Carpodacus thura	x			
Propyrrhula subhimachala		o x		
Haematospiza sipahi	x			
Pyrrhula nipalensis	x			

The valley is an important divide for bird as well as plant species as first pointed out by Fleming *et al.* (1984). Approximately 436 breeding forest bird species have been found in eastern forests and 338 in the west. Western forests are poorly recorded compared to those in the east. There are 37 breeding species which have been located east of the Kali Gandaki and also to the west of Nepal in India (Table 9). Assuming there is suitable habitat they could all be found in west Nepal, but even if all of them do occur, eastern forests are still considerably richer. Eastern forests also support a much larger number of breeding species for which Nepal may hold significant populations (109 species) compared to those in the west (68 species).

The apparent world ranges of 75 breeding Nepalese forest species lie east of the valley including 38 Himalayan species which reach the western limit of their ranges either in the valley or its watershed (Table 9). The valley is a significant, but less important eastern barrier to species occurring in the west. The world ranges of eight species which largely favour coniferous forests lie west of the valley, three of them occurring east to the river (Table 9). The Nepalese ranges of Bar-tailed Treecreeper *Certhia himalayana*, White-cheeked Nuthatch *Sitta leucopsis* and Chestnut-eared Bunting *Emberiza fucata* cease at the Kali Gandaki and that of Koklass Pheasant *Pucrasia macrolopha*, Stoliczka's Tit-Warbler *Leptopoecile sophiae* and Red-fronted Serin *Serinus pusillus* extend just to the east of it, but other races of all of them occur much further east in China and Burma.

Another very deep river valley, the Arun, in east Nepal also marks a change in avifauna, but is less important than that of the Kali Gandaki. Rainfall in the Arun valley and eastwards is higher than in most of the rest of the country and wet forests of evergreens, rhododendrons and mixed broadleaves are much more widespread. There are 20 species whose range in Nepal is apparently confined to this area (Table 9).

Forests to the south of Annapurna and Himal Chuli in central Nepal are similar to those in parts of the far east. This is the wettest part of the country as the monsoon rain from India has relatively low hills to cross before reaching it. Four species are restricted to these particularly wet forests of both central and east Nepal: Rusty-fronted Barwing *Actinodura egertoni* (in subtropical and lower temperate forests), Golden Babbler *Stachyris chrysaea* (in lower temperate forests), and Brown Parrotbill *Paradoxornis unicolor* and Golden-breasted Fulvetta *Alcippe chrysotis* (in upper temperate forests).

FOREST SPECIES ADAPTED TO MAN-MODIFIED HABITATS

There are 76 species, only 16 percent of all forest birds which have adapted to breed in habitats heavily modified or created by man, such as groves, gardens, scrub, and trees and bushes at the edges of cultivation (see Appendix). Once they presumably bred in forest edges and clearings. Forest losses must have benefited them, notably the Spiny Babbler *Turdoides nipalensis*, perhaps Nepal's only endemic bird which inhabits dense 'shrubberies' in the subtropical and lower temperate zones. It has been recorded throughout the country, almost to the eastern and western borders and may well also occur in India. The Spiny Babbler is easily overlooked because of its skulking habitats, but in fact is fairly common in some places, such as the hills surrounding the Kathmandu valley (Proud 1959). All of the species in this group are common or fairly common with the exception of seven terai birds which are common in the Gangetic plain and the Red-necked Falcon *Falco chicquera* which is declining for unknown reasons (Inskipp and Inskipp 1985). A number of other species commonly feed or pass through secondary habitats, but are not known to breed there.

ANALYSIS OF FOREST TYPES AND THEIR BIRD COMMUNITIES

TROPICAL ZONE

Forests

Sal *Shorea robusta* forest

Forest where sal is dominant covers by far the greater part of the zone and its composition varies remarkably little from east to west throughout Nepal. Epiphytes and climbers are scarce. There is thick undergrowth in a few forests, but many are suffering badly from regular and often annual burning and over-grazing and have a floor covered only with dry leaves. Sal forests in the hills and plains differ in appearance. Those in the plains have taller trees, generally about 24 m in height and occasionally reaching 47 m and with a denser understorey. Hill sal forests have trees only 12-16 m high and a greater abundance of grasses in the ground layer.

Acacia catechu-Dalbergia sissoo forest

Grows in narrow strips on newly deposited alluvium along streams and rivers and is surrounded by sal forest.

Tropical evergreen forest

Grows in narrow belts, usually only 100 m wide in sal forest, either along water courses or in gulleys and often in shady north-facing sites. In some areas the trees are up to 47 m high with an understorey of smaller trees, but few shrubs or climbers. Other forests are only about 19 m high and have a dense growth of shrubs, climbers, bamboos and palms. Tropical evergreen forest in the south-east is considerably more widespread and richer in plant species than at Chitwan in east central Nepal and much richer than in the west where it is scarce. Over-exploitation for fuelwood, fodder, bamboo and over-grazing are severely threatening many forests. It has been suggested that this forest type may be restricted to damp shady sites by the regular fires in the terai (Stainton 1972).

Terminalia-Anogeissus deciduous hill forest

Occurs extensively in the foothills of western Nepal on south faces. Elsewhere it is confined to dry south-facing slopes in the larger river valleys.

Birds
Data for this section are given in Table 10.

Page 44

Carol Inskipp

Table 10: Analysis of forest birds and their breeding habitats.

Column 1: Breeding species
Column 2: Species which may have internationally significant populations in Nepal
Column 3: Species which may have internationally significant populations in Nepal and for which the country may be especially important
Column 4: Species considered at risk in Nepal
Column 5: Species restricted to the zone
Column 6: Species restricted to the Kali Gandaki valley and further east
Column 7: Species restricted to the Kali Gandaki valley and further west
Column 8: Species adapted to man-modified habitats

	Number of species							
	1	2	3	4	5	6	7	8
Tropical	216	7	3	55	86	24	0	59
East	151	4	2	44				
Central	152	6	2	44				
Far west	121	5	2	30				
Subtropical	200	17	5	42	11	18	0	53
Schima-Castanopsis	144	13	3	39				
West (all forest types)	103	10	1	5				
Lower temperate	172	42	8	34	16	32	5	28
Mixed broadleaves	119	33	5	33				
Quercus lamellosa	112	31	3	26				
Quercus lanata	71	15	0	1				
West (all forest types)	97	26	2	2				
Upper temperate	158	65	10	29	14	24	8	10
Mixed broadleaves	98	42	5	22				
Quercus semecarpifolia	97	41	5	24				
West (all forest types)	105	42	5	17				
West (conifers only)	52	18	1	4				
Subalpine	103	58	13	7	29	13	5	5
Abies spectabilis	40	21	1	4				
Betula utilis	33	24	3	3				
Rhododendron spp.	27	22	4	7				
Juniperus spp.	27	15	3	6				
Alpine	19	5	0	0	9	0	0	0

Note: Numbers all refer to breeding or probably breeding species.

The tropical forests are the richest for birds in Nepal, with a total of 204 breeding species.   An additional 12 species were recorded by Hodgson last century, but five of them may have originated outside of Nepal (see p.37).

A high proportion (42 percent) of species are restricted to the tropical zone compared to the other zones.   Nearly all tropical species (90 percent) are sedentary residents and the rest are summer migrants.

The zone has very few species which may have significant breeding populations in the country.   This is to be expected as Nepal's tropical forests are of relatively small extent compared to those further south and east in Asia.

Species adapted to breed in scrub, edges of cultivation, groves or gardens comprise only 29 percent of all those in the zone.   The large majority (82 percent) of Nepal's forest birds which have adapted to these habitats are tropical species.

As the sal forests have been subject to burning so extensively it is probably impossible to determine the natural composition of their bird communities except within the protected areas of the Royal Chitwan National Park (Chitwan) and the Royal Bardia Wildlife Reserve (Bardia).   Three main tropical forest types for birds can be distinguished: eastern (lying between the Mechi and Kosi rivers), central (lying between Butwal and Hetaura and including Chitwan) and western (in Royal Sukla Phanta and Bardia Reserves).

The tropical evergreen forest is of great importance although, because it only occurs in narrow strips within the sal, a distinct bird community could not be identified.   However, 28 species are only associated with this forest type and seven of them are confined to the east probably because the tropical evergreen forest is more extensive and varied there (Table 11).

Table 11: Breeding birds associated with tropical evergreen forest.

	Black Baza *Aviceda leuphotes*
	Rufous-bellied Eagle *Hieraaetus kienerii*
	Pompadour Green Pigeon *Treron pompadora*
	Thick-billed Green Pigeon *Treron curvirostra*
	Vernal Hanging Parrot *Loriculus vernalis*
+	Ruddy Kingfisher *Halcyon coromanda*
* +	Blyth's Kingfisher *Alcedo hercules*
	Deep-blue Kingfisher *Alcedo meninting*
* +	Blue-eared Barbet *Megalaima australis*
	White-browed Piculet *Sasia ochracea*
* +	Pale-headed Woodpecker *Gecinulus grantia*
	Long-tailed Broadbill *Psarisomus dalhousiae*
	Large Woodshrike *Tephrodornis gularis*
	White-throated Bulbul *Criniger flaveolus*
+	Asian Fairy Bluebird *Irena puella*
* +	Rufous-faced Warbler *Abroscopus albogularis*
	Yellow-bellied Warbler *Abroscopus superciliaris*
	Pale-chinned Flycatcher *Cyornis poliogenys*
*	Abbott's Babbler *Trichastoma abbotti*
	Grey-throated Babbler *Stachyris nigriceps*
	Lesser Necklaced Laughing-thrush *Garrulax monileger*
	Greater Necklaced Laughing-thrush *Garrulax pectoralis*
* +	White-hooded Babbler *Gampsorhynchus rufulus*

Table 11 (contd)

> + Sultan Tit *Melanochlora sultanea*
>   Ruby-cheeked Sunbird *Anthrapetes singalensis*
>   Streaked Spiderhunter *Arachnothera magna*
>   Little Spiderhunter *Arachnothera longirostra*
> * + Yellow-vented Flowerpecker *Dicaeum chrysorrheum*

> * indicates species confined to eastern forests
> + indicates species only recorded outside present protected areas

A total of 21 species extend to central forests. although six of them: Black Baza *Aviceda leuphotes*, White-throated Bulbul *Criniger flaveolus*, Long-tailed Broadbill *Psarisomus dalhousiae*, Asian Fairy Bluebird *Irena puella*, Ruby-cheeked Sunbird *Anthreptes singalensis* and Little Spiderhunter *Arachnothera longirostra* have been recorded much more frequently in eastern than in central forests.

The eastern forests are likely to be the richest because of their additional evergreen forest avifauna. The number of species found so far is almost the same as the total in central forests (151 species), but the latter are much better recorded. Although Chitwan is the only known locality in Nepal for Rufous-necked Laughing-thrush *Garrulax ruficollis*, a species for which the country may be important, the bird could well occur further east. The endangered Grey-headed Fishing Eagle *Ichthyophaga ichthyaetus* and Lesser Fishing Eagle *I. nana* breed at Chitwan, but probably do not do so in eastern forests, because of the lack of lakes and large rivers near wooded country.

Western forests are notably less diverse (121 species) than those further east (152 species), even counting the additional birds which could be present (Table 9). There are four species: Brown-headed Barbet *Megalaima zeylanica*, White-naped Woodpecker *Chrysocolaptes festivus*, Jungle Prinia *Prinia sylvatica* and Tickell's Blue Flycatcher *Cyornis tickelliae* which favour open dry forests and are limited to the west, although all occur further east in India (Inskipp and Inskipp 1985). Nepal's western forests are also important for the endangered Great Slaty Woodpecker *Mulleripicus pulverulentus* whose main population is in the far west, although it does occur east to Chitwan.

A high proportion of tropical forest species, 55 (27 percent of the total), are considered to be at risk mainly because of forest clearance and damage. Many of them, especially those dependent on evergreen forests, are very locally distributed.

SUBTROPICAL ZONE

Lies between approximately 1,000 m and 2,000 m in the west and 1,000 m and 1,700 m in the east.

Forests

Schima-Castanopsis forests

Schima-Castanopsis once covered much of the subtropical zone in central and eastern Nepal. In the Arun valley and eastwards, *Schima wallichii/Castanopsis indica* was commonest below 1,200 m and *Schima wallichii/Castanopsis tribuloides* above this altitude. Most of these *Schima-Castanopsis* forests have been replaced by intensive cultivation. In many places only small patches remain and these have often been damaged by removal of wood and foliage, burning and by over-grazing.

Pinus roxburghii forest

There are extensive pine forests on all aspects in west Nepal, but in the centre and east they are confined to drier situations, such as on southern slopes and on the lower slopes of large river valleys. *Pinus roxburghii* is absent in wet areas such as that immediately to the south of Annapurna and in the wetter parts of the east. The forests are typically open and composed of old pines with no understorey because of frequent fires.

Alnus nepalensis forest

Grows along streams and in ravines and often colonises abandoned cultivation and soil exposed by landslips. It grows within the above two forest types.

Riverine forest with *Toona* and *Albizia* spp.

A wet subtropical forest which grows along water courses and in ravines within *Schima-Castanopsis* forest. It often exists as narrow strips in country where the surrounding forest has been cleared for cultivation.

Subtropical evergreen forest

Occurs in the foothills of eastern Nepal between the Kosi and Mechi rivers. It is of limited extent and is being exploited in a similar way to tropical evergreen forests.

Birds

Data for this secton are given in Table 10.

Subtropical forests support a wide variety of birds totalling 183 breeding or species with perhaps another 17 species not recorded since the last century; seven of these latter species might possibly have never occurred in Nepal (see p.37).

A small proportion of subtropical species (6 percent) are confined to this zone in the breeding season.

Nepal may hold significant breeding populations of comparatively few subtropical species. This is not surprising, as in common with tropical forests, much more extensive subtropical forests occur elsewhere in Asia.

The relatively large proportion of 29 percent of those in the zone, have adapted to breed in habitats heavily modified or created by man.

The subtropical forests can be divided into three main types: wet broadleaved forests in the Mai valley area in the far east, other broadleaved forests east of the Kali Gandaki valley and forests west of the valley. All subtropical species have been found in forests in the Kali Gandaki valley and further east. The western limits of 18 species lie in this region (Table 9).

The *Schima-Castanopsis* forests together with the riverine forests within them and the relatively small areas of subtropical evergreen forests are by far the most important habitats for birds in the zone. Those in the wetter parts of the country such as the Arun valley, the area immediately south of Annapurna, the Mai valley area in the far east and on Phulchowki mountain in the Kathmandu valley are the most diverse. Many subtropical species at risk (70 percent) and 56 percent of those which may have significant populations in Nepal prefer or require damp broadleaved or evergreen forests. The importance for birds of wet forests in the far east is described on p.71. Three species are only known from there at present: Chestnut-backed Sibia *Heterophasia annectans* in the Mai valley, Hill Prinia *Prinia atrogularis* in the Mewa Khola and Mai valleys and on Hans Pokhari Danda, and White-naped Yuhina *Yuhina bakeri* on Hans Pokhari Danda.

No significant difference could be found between the bird communities of the *Schima wallichii/Castanopsis indica* and *Schima wallichii/Castanopsis tribuloides* forests. There is very little information available on birds of the *Alnus nepalensis* forests, but Cronin (1979) describes them as very species-poor for birds as well as for other wildlife.

Subtropical forests west of the Kali Gandaki are under recorded. Even when allowance is made for the occurrence of other likely species (assuming suitable habitat is available for them, and in many cases this is doubtful), western forests still have a much less diverse avifauna than those in the east (Table 9). Although pine forests cover much of the zone the limited habitat information recorded indicates that many birds have been found in strips of broadleaved forests in gulleys or along streams. Only 14 species so far recorded in the west have been described as frequenting pine forest. At least 48 other species located are unlikely to occur in pines, especially as these forests have little or no understorey (Ali and Ripley 1984; Fleming *et al.* 1984).

As many as 23 percent of Nepal's subtropical forest birds are threatened and half of these are thought to be endangered.

LOWER TEMPERATE ZONE

Lies between approximately 2,000 m and 2,700 m in the west and 1,700 m and 2,400 m in the east.

Forests

Quercus lamellosa forest

A wet forest growing between 1,900 m and 2,600 m which is especially common in areas of high rainfall such as the southern slopes of the Annapurna Himal and Himal Chuli Himal and the upper Arun and Tamur valleys. Elsewhere it is confined to north and west faces. The *Lauraceae* of the lower temperate mixed broadleaved forest are often associated with it. There is usually a dense understorey and the trees are clothed with mosses, ferns and epiphytes. The upper limit of cultivation in the wet areas where *Quercus lamellosa* grows is about 2,100 m and so the lower parts of the forest have frequently been cleared or over-exploited.

Lower temperate mixed broadleaved forest with abundant *Lauraceae*

Occurs in the well-cultivated altitudinal range between 1,500 m and 2,100 m and so much of this forest has been cleared. It is a mainly evergreen forest which grows in the wetter parts of Nepal, usually on north or west faces. Like the previous forest type it is particularly abundant on the southern slopes of Annapurna and Himal Chuli and in the upper Arun and Tamur valleys. A number of *Lauraceae* species are present and sometimes *Quercus lamellosa*.

Quercus leucotrichophora and *Q. lanata*

A dry form of oak forest found between 1,750 m and 2,400 m. It grows on all aspects in the west. In the centre and east it mainly occurs on southern slopes and the sides of the larger river valleys and is absent from the very high rainfall areas. Even in the less populated west these forests are under heavy pressure from human activities and are limited to slopes which are too steep to climb and in areas which are farther than a day's journey from a village (Shrestha 1984-1985). Elsewhere the forest has been largely cleared altogether or much damaged by removal of wood and foliage, over-grazing and burning. Stainton (1972) stated 'it can be seen in its natural form only in a few places on the outlying spurs of the main ranges'. Forests in the centre and east are open with little undergrowth as a result of exploitation. Stainton (1972) describes *Quercus lanata* as the least botanically interesting of all Nepal's forests.

Quercus floribunda forest

This is confined to western Nepal between 2,100 m and 2,850 m. It replaces the previous type on wet sites, mainly on north or west faces. *Aesculus indica* and *Acer* spp. are often mixed with it.

Pinus wallichiana (lower type)

This species has a large altitudinal range, from 1,800 m to 4,000 m. It colonises abandoned fields and grazings and in the midlands it probably only forms forest on land modified by man. Here the pine forest is very open, often with scrub oak amongst it. Extensive *Pinus wallichiana* forest occurs in west Nepal, especially on dry southern faces. Mature stands of pine can be found here, but these are much less common further east, probably because of the greater demands for wood.

Birds

Data for this section are given in Table 10.

There are 167 breeding species in the lower temperate zone with perhaps another five collected in the last century if they really were obtained in Nepal.

Only 10 percent of lower temperate forest species are restricted to this zone.

Nepal may have significant breeding populations of as many as 42 of the species (25 percent of the total).

Only 17 percent of the species breed in habitats highly modified by people.

Forests to the east and west of the Kali Gandaki valley are important for birds, but eastern forests are richer, even if the additional species which could occur are counted (Table 9). Eastern forests also support a larger number of species for which Nepal may hold significant populations. There are 32 species whose ranges lie east of the Kali Gandaki and Nepal may hold significant populations of 14 of these (Tables 6 and 9). Five reach the eastern limits of their ranges at the river and Nepal may be important for three of these, notably the Cheer Pheasant *Catreus wallichii*, which is listed as 'Endangered' in the Red Data Book (King 1978-1979).

Five main forest types for birds can be identified in the zone and include wet forests of the Arun valley and eastwards (mainly mixed broadleaved and some *Quercus lamellosa*), wet forests lying between the Kali Gandaki and Arun valleys (mainly *Q. lamellosa* with some mixed broadleaved), dry oak forests of the centre and east (*Q. lanata*) and forests west of the Kali Gandaki (*Q. leucotrichophora* and *Q. lanata*, *Q. floribunda* and *Pinus wallichiana*). Wet forests in the Mai valley area and perhaps elsewhere in far east Nepal are sufficiently different to be considered a separate type. The importance of this area for birds is described on p.71. Three species are known only from there at present: Chestnut-backed Sibia *Heterophasia annectans* and Black-spotted Yellow Tit *Parus spilonotus* in the Mai valley and Hill Prinia *Prinia atrogularis* from there, the Mewa Khola valley and Hans Pokhari Danda.

Bird communities of the mixed broadleaved and *Quercus lamellosa* forests could not be completely separated using available information, mainly because the forests are often intermixed. The wet forests of the Arun valley and eastwards are the richest for birds with 119 breeding species recorded, 88 percent of those which occur in the zone. These include the high number of 33 species for which Nepal may hold significant populations and five for which the country may be especially important.

The avifauna of the wet central forests is almost as diverse (112 species) and is similar in composition to the previous type, but nine eastern species have not been found, probably because the mixed broadleaved forest is not so extensive as in the east. These are Slaty-bellied Tesia *Tesia olivea*, Broad-billed Warbler *Abroscopus hodgsoni*, Yellow-bellied Warbler *A. superciliaris*, Sapphire Flycatcher *Ficedula sapphira*, Chestnut-backed Sibia *Heterophasia annectans* and Black-spotted Tit *Parus spilonotus*. Two additional species occur: Lanceolated Jay *Garrulus lanceolatus* (whose range terminates west of the Arun valley) and Cinnamon Sparrow *Passer rutilans*.

The *Quercus lanata* forests of the centre and east are impoverished compared to the previous two forest types. Only 71 breeding species have been located and just four species which have not been found in the other two types as they prefer drier and more open forests. These are Eurasian Hobby *Falco subbuteo*, Brown-fronted Pied Woodpecker *Dendrocopos auriceps*, Asian Magpie Robin *Copsychus saularis* and Thick-billed Flowerpecker *Dicaeum agile*. This low species diversity can be partly attributed to the lack of good stands of *Quercus lanata* remaining east of the Kali Gandaki. All available records are from degraded forests. Indeed it is probably no longer possible to determine the composition of the *Q. lanata* bird community in central and eastern Nepal as the undamaged forest is reduced to isolated small remnants. However, even the intact *Q. lanata* forest must have been unsuitable for the many lower temperate forest species (approximately 28) which require damp conditions with luxuriant undergrowth and occur in *Q. lamellosa*. Some *Q. lanata* forests west of the Kali Gandaki are still in good condition and need further investigation. *Pinus wallichiana* forests in the centre and east are very species-poor.

Data on lower temperate species in western Nepal are relatively limited. A total of 96 breeding species has so far been reported from all western lower temperate forests. Nepal may support significant populations of 26 of these. Most species prefer broadleaved forests with the notable exceptions of four of the five species confined to the west.

The high proportion at 20 percent of lower temperate forest species is at risk.

UPPER TEMPERATE ZONE

Occurs between about 2,700 m and 3,100 m in the west and centre and from 2,400 m to 2,800 m in the east.

Forests

Upper temperate forests are much less disturbed than those lower down, especially those in the west, because they mainly lie above the limit of cultivation. However, they are still used for grazing, especially in the summer months, often by herders on their way to pastures higher up, and are exploited to some extent for wood and foliage.

Quercus semecarpifolia forest

In the centre and east it grows between 2,400 m and 3,400 m, but in the west it extends much higher, up to the tree line at 3,700 m in the Karnali region. It is more common on south-facing slopes. Where the forest is not burned in central and east Nepal, the trees are often covered with mosses, ferns and epiphytes. Sometimes there may be overlap between this forest and upper temperate broadleaved forest.

Upper temperate mixed broadleaved forest

A wet forest found east of the Kali Gandaki river between 2,400 m and 3,150 m, mainly on north and west faces. It is very mixed in composition with several *Acer* and *Lauraceae* species as well as *Rhododendron arboreum* in the understorey. *Tsuga dumosa* often occurs and may form almost pure patches on ridges and drier sites. This forest type and rhododendron forest share many component species and forest intermediate between the two types occurs.

Rhododendron spp. forest

This is forest where rhododendrons dominate the upper storey and any other tree species which may be present are only irregularly scattered amongst them. Rhododendron species are also common in forests of *Abies*, *Betula*, *Tsuga* and upper temperate mixed broadleaved forest, but they are part of the understorey beneath the other trees. Rhododendrons are most widespread and species diversity is greater in very high rainfall areas such as in the upper Arun and Tamur valleys.

Upper temperate coniferous forest

Pinus wallichiana, which occurs up to 4,000 m, is distributed nearly throughout and is often almost pure, especially on south faces (see Lower Temperate Forests, p.48). In moister areas in western Nepal, especially in the Humla-Jumla area it is associated with *Abies pindrow*, *Picea smithiana* and very locally with *Cedrus deodara*. In the very dry Mustang region it is associated with *Juniperus indica* and in central and eastern Nepal with *Tsuga dumosa*, *Taxus baccata* and *Acer* spp. Both *Picea smithiana* and *Abies pindrow* sometimes form almost pure stands and have been distinguished as separate forest types (Stainton 1972). Here they are included in the upper temperate coniferous forest as they nearly always have a few *Pinus wallichiana* amongst them.

Arundinaria spp. and *Bambusa* spp.

Bamboo flourishes in very high rainfall areas and in a few places such as in the Modi Khola valley, south of Annapurna, it forms pure dense stands up to 7m high. *Arundinaria* spp. and *Bambusa* spp. are also common in the understorey of *Quercus semecarpifolia*, upper temperate mixed broadleaved forests and rhododendron forests.

Birds

Data for this section are given in Table 10.

The most internationally important of all Nepal's forests for breeding birds are those in the upper temperate and subalpine zones. As many as 65 upper temperate forest species (41 percent of the total) may have significant populations in Nepal and the country may be especially important for ten of them. There are 158 breeding birds and only 9 percent of them are confined to the zone. Long-billed Wren-Babbler *Rimator malacoptilus* may have been collected in the last century (see p.37) and could occur in the wetter eastern forests. Only ten forest species breed in habitats heavily modified by people. Both forests to the east and west of the Kali Gandaki are of high importance, although eastern forests are more diverse, even when an allowance is made for the possible occurrence of other species in the west (Table 9). The Kali Gandaki valley forms the western limit of the ranges of 24 species. Nepal may support significant breeding populations of 15 of these and be especially important for five of them (Tables 6 and 9). Another eight species reach the eastern limit of their world ranges at the river and White-cheeked Nuthatch *Sitta leucopsis* does not occur further east in Nepal. Half of these species may have significant breeding populations in Nepal and the country may be especially important for two of them, including the Cheer Pheasant *Catreus wallichii*.

Four main forest types for birds can be identified in the upper temperate zone: eastern broadleaved forests, from the Arun eastwards (mainly mixed broadleaves, rhododendron and some *Quercus semecarpifolia*, central broadleaved forests between the Kali Gandaki and Arun valleys (mainly *Q. semecarpifolia* with some mixed broadleaves), western broadleaved forests (*Q. semecarpifolia*) and western coniferous forests.

Coniferous forest of the centre and east has not been treated as a separate type as the bird community is much poorer (30 species) than in western coniferous forests (52 species). Only one additional species, Spot-winged Grosbeak *Mycerobas melanozanthos* has been found and it is also likely to occur in the west.

The bird communities of the mixed broadleaved and *Quercus semecarpifolia* cannot be clearly distinguished using available data, mainly because the forests are often intermixed. The eastern broadleaved forests have the most diverse avifauna with 98 species recorded including 42 species (43 percent of the total) for which Nepal may support significant breeding populations and five for which the country may be especially important. More species are likely to be found with further study. Bird communities of central broadleaved forests are very similar apart from the addition of Red-browed Finch *Callacanthis burtoni* and the absence of Sapphire Flycatcher *Ficedula sapphira* and Rufous-capped Babbler *Stachyris ruficeps*.

Bamboo *Arundinaria* spp. and *Bambusa* spp. is an important component of upper temperate vegetation for many birds especially Satyr Tragopan *Tragopan satyra*, Grey-cheeked Warbler *Seicercus poliogenys*, Snowy-browed Flycatcher *Ficedula hyperythra*, Slender-billed Scimitar-Babbler *Xiphirhynchus superciliaris*, Rufous-capped Babbler *Stachyris ruficeps*, Black-throated Parrotbill *Paradoxornis nipalensis* and Scaly Laughing-thrush *Garrulax subunicolor*. Pure bamboo stands are very impoverished for birds but four are mainly confined to this habitat: Great Parrotbill *Conostoma aemodium*, Brown Parrotbill *Paradoxornis unicolor*, Fulvous Parrotbill *P. fulvifrons* and Golden-breasted Fulvetta *Alcippe chrysotis*. Nepal may be of special importance for all of these species.

Western Nepal forests have been little studied. A total of 42 species for which Nepal may hold significant populations and five for which the country may be especially important have been found. Habitat information on the 105 breeding species recorded is limited. However, the coniferous forests are important even though they are species-poor. Nepal may have significant populations of 35 percent of the species recorded there so far. Another 18 which are likely to occur in conifers have been recorded in the western upper temperate zone. Eight of the nine species restricted to the west require coniferous forests.

A total of 76 species has been so far located in western broadleaved forests. As these western forests are drier, more open and have fewer epiphytes, bird species diversity is likely to be lower than that in eastern forests of the same type. Nepal may have significant populations for 40 percent of the species recorded and may be especially important for five of them, including Cheer Pheasant *Catreus wallichii*, the only species from this forest type which is restricted to the west.

Mixed conifer/broadleaved forests are important for birds in the upper temperate zone, although nearly all species also often occur in one or more of the other types. There are 15 upper temperate forest species which cannot be assigned to the above main forest types, notably Large-billed Leaf Warbler *Phylloscopus magnirostris* (occurs in dense vegetation along streams), Spotted Laughing-thrush *Garrulax ocellatus* (found in mixed conifer/oak and bamboo forests), Fire-capped Tit *Cephalopyrus flammiceps* (recorded in maple *Acer* spp. and mixed maple/conifer forest), Common Crossbill *Loxia curvirostra* associated with hemlock *Tsuga dumosa* and Spot-winged Grosbeak *Mycerobas melanozanthos* (favours hemlock *T. dumosa* and maple/conifer forest). Approximately 25 of all upper temperate forest species are threatened.

SUBALPINE ZONE

Lies between 3,000 m and 4,200 m in the west, and 3,000 m to 3,800 m in the east.

Forests

Some of the least-disturbed forests occur in this zone, especially those of *Abies spectabilis* (Kenting 1986). Shrestha (1984-1985) has described those in the Karnali zone in the west as consisting of a good deal of undisturbed natural forests. However, locally, such as in the Sagarmatha National Park, collection of wood for tourist trekkers and mountaineers is a severe threat (see p.25), especially as high altitude forests are much slower-growing than those lower down (Mishra 1986).

Abies spectabilis forest

Abies spectabilis forest usually forms a continuous belt between 3,000 m and 3,500 m on the southern side of the main ranges in central Nepal. This forest normally has a dense understorey of rhododendrons. Sometimes the fir grows up to the tree line, but is usually superseded by *Betula utilis* forest above 3,500 m. *Larix griffithiana* and *L. himalaica* are locally common within it, but rarely form pure stands. *Abies* forest is less widespread in the east and in very high rainfall areas it is largely replaced by rhododendron forest where *Abies* only occurs singly or in small groups.   In western Nepal, *Abies spectabilis* is often mixed with *Quercus semecarpifolia* and *Betula utilis*.

Betula utilis forest

Found between 3,300 m and the tree line.   In central Nepal it usually occurs above the *Abies spectabilis* belt and has an understorey of *Rhododendron* spp.   It mainly grows as scattered trees among rhododendrons in the east and is frequently mixed with *Abies* spp. and *Quercus semecarpifolia* in the west.

Rhododendron spp. forest

As in the upper temperate zone rhododendron forest often replaces other forest types on very wet sites in eastern Nepal.   A large number of rhododendron species occurs and occasional trees of *Betula utilis*, *Abies spectabilis*, *Tsuga dumosa* and species typical of the upper temperate mixed broadleaved forest are sometimes found among them.

Juniperus spp. forest

Juniperus indica occurs in drier areas, both as a tree and a small shrub.   In some places it forms pure forest.   *J. recurva* occurs as a component of the subalpine forest in the wetter parts of the midlands and is often mixed with other trees.

Arundinaria spp. and *Bambusa* spp.

Arundinaria spp. and *Bambusa* spp. are common in the understorey of high rainfall areas and sometimes form pure stands.

Birds

Data for this section are given in Table 10.

Nepal's subalpine forests are of great international importance for breeding birds, although they are poorer in species (103) than zones at lower altitudes. The country may hold significant breeding populations of over half the bird species in the zone (57 percent) and may be especially important for 13 of them. The high proportion of 28 percent of species are restricted to subalpine forests and only five species are known to breed in habitats heavily modified by people.

Although subalpine forests throughout the country are valuable, forests to the east of the Kali Gandaki valley have a more diverse avifauna and more species for which Nepal may be important than forests to the west (Tables 6 and 9).   The western ranges of 13 species terminate at the valley and Nepal may have significant populations of ten of these.   The ranges of another five species are limited to west of the Kali Gandaki and four of them may have significant populations in the country.

Although forest mixtures occur widely, four main types for birds can be distinguished: *Abies spectabilis*, *Betula utilis*, *Rhododendron* spp. and *Juniperus* spp.   Available habitat information for breeding birds is limited, but the numbers of breeding birds which may have significant populations in Nepal are high for all forest types.   Bamboo is especially valuable in the subalpine forests and although the pure stands support very few species, Coral-billed Scimitar-Babbler *Pomatorhinus ferruginosus*, Great Parrotbill *Conostoma aemodium* and Fulvous Parrotbill *Paradoxornis fulvifrons* are almost entirely confined to them.   Nepal may be of special importance for both parrotbills.

About 7 percent of subalpine forest species are thought to be threatened.

ALPINE ZONE

Lies between the tree-line (3,800 m in the east and 4,200 m in the west) and the region of perpetual snow.

Forests

In the same way as subalpine forests (see p.25), forests in a few parts of the alpine zone are dangerously threatened by excessive collection of firewood for tourists and particularly for mountaineering expeditions (Mishra 1986). One such example is at Makalu Base Camp in the Barun valley where the forests are otherwise considered 'the most pristine and diverse jungle habitat remaining in Nepal' (Shrestha *et al*. 1985).

Alpine scrub grows above the tree line. Rhododendrons, junipers and some other shrubby species, such as *Hippophae rhamnoides* and *Cotoneaster microphyllus* occur up to 4,870 m. Rhododendrons are abundant in east Nepal and junipers are common in the west. In north-west Nepal, north of the Dhaulagiri-Annapurna massif there is extensive steppe country (Dolpo, Mustang and Manang districts) with vegetation similar to that of the Tibetan plateau. This is dominated by *Caragana*, a low spiny shrub, which is usually less than 1.5 m in height and grows up to an altitude of 5,200 m.

Birds

Data for this section are given in Table 10.

Species diversity is very low. Breeding birds total 19, of which five may have significant breeding populations in Nepal. Nearly all are common in the northern part of the country in the Tibetan steppe and none is considered to be at risk.

DISTRIBUTIONAL TRENDS

The following significant and interesting trends occur from the tropical to the alpine zones (Table 10 and Figures 6, 7 and 8):

(a) the number of breeding species decreases;
(b) the number of breeding species which may have internationally significant populations in Nepal increases up to the subalpine zone;
(c) the number of breeding species at risk decreases.

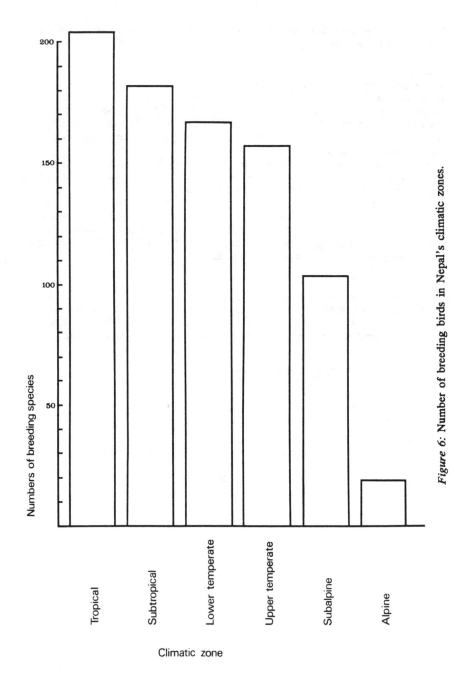

Figure 6: Number of breeding birds in Nepal's climatic zones.

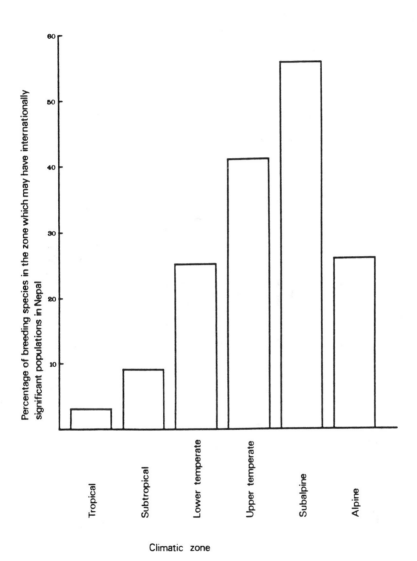

Figure 7: Percentage of breeding birds in Nepal's climatic zones which may have internationally significant populations in the country.

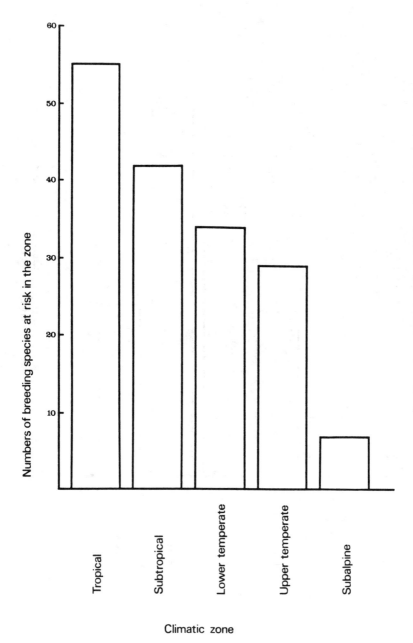

Figure 8: Number of breeding birds at risk in Nepal's climatic zones.

Cheer Pheasant *Catreus wallichii* (Dave Showler)

BIRDS AT RISK

Table 12 lists breeding birds species that are considered to be at risk in Nepal, and summarises their habitats, ways in which they are threatened, and occurrence outside reserves.

There are 131 altogether (22 percent of the total) and 110 of them (84 percent) are dependent on forests. The two fishing eagles require both forests and wetlands as they breed in forests close to large rivers or lakes. The remaining species inhabit grasslands (12), wetlands (5), scrub (1), open country (1), bare stony ground (1); and the habitat of one is unknown. The low proportion of species which breed outside forests is to be expected as natural grasslands and wetlands are of relatively small extent in Nepal. Apart from the scrub species, all of the non-forest birds occur in the tropical zone.

Habitat loss is the major threat to 95 percent of the birds at risk. The spread of cultivation has reduced the extent of grasslands and now almost all lie within protected forest areas (Inskipp and Inskipp 1983). Pools and marshes which must have once existed throughout the lowlands have now been largely drained to make way for agriculture. These losses were partly compensated for by the large expanse of open water, marshes and reedbeds created by the construction of Kosi Barrage in 1964. Forest losses and deterioration are described on p.25.

Three-quarters of threatened forest birds (84 altogether, Table 12) require dense or moist conditions with plenty of undergrowth or trees covered in epiphytes. These species are particularly sensitive to changes in the forest ecosystem. Forest birds mainly come from the tropical, subtropical or lower temperate zones where forests have been most severely depleted. A total of 58 species breeds in tropical forests and 57 in subtropical and/or lower temperate forests, with 30 species confined to the former and 21 to the latter two zones.

Some birds have specialist needs and are very locally distributed. The Pale-headed Woodpecker *Gecinulus grantia*, Yellow-bellied Bush-Warbler *Cettia acanthizoides*, Coral-billed Scimitar-Babbler *Pomatorhinus ferruginosus*, Great Parrotbill *Conostoma aemodium*, Brown Parrotbill *Paradoxornis unicolor*, Fulvous Parrotbill *P. fulvifrons* and Golden-breasted Fulvetta *Alcippe chrysotis* occur mainly in pure bamboo stands. A number of other species favour forests with a bamboo understorey, notably Satyr Tragopan *Tragopan satyra*, White-browed Piculet *Sasia ochracea*, White-tailed Robin *Cinclidium leucurum*, Grey-cheeked Warbler *Seicercus poliogenys*, Rufous-faced Warbler *Abroscopus albogularis*, Slender-billed Scimitar-Babbler *Xiphirhynchus superciliaris*, Golden Babbler *Stachyris chrysaea*, Black-throated Parrotbill *Paradoxornis nipalensis*, Rufous-necked Laughing-thrush *Garrulax ruficollis* and Fire-tailed Myzornis *Myzornis pyrrhoura*. All of these must be threatened to some degree by bamboo losses (described on p.29).

The Great Slaty Woodpecker *Mulleripicus pulverulentus* frequents mature sal forest in the terai and is threatened by tree-felling. The Indian Courser *Cursorius coromandelicus* inhabits stony ground and dry open areas in the terai. This marginal land must have always been of very limited extent in Nepal, but now

Table 12: Breeding birds at risk in Nepal (all species).

Categories of threat are those defined by the International Union for Conservation of Nature and Natural Resources (Collar and Stuart 1985).

E   Endangered: taxa in danger of extinction in Nepal and whose survival is unlikely if the causal factors continue operating

V   Vulnerable: taxa believed to move into Endangered category in the near future if the causal factors continue operating

R   Rare: taxa with small Nepalese populations that are not at present Endangered or Vulnerable, but are at risk

I   Indeterminate: taxa known to be Endangered, Vulnerable or Rare in Nepal, but where there is not enough information to say which of the three categories is appropriate

Habitats

F   Forest (see Appendix for forest types)
G   Grassland
W   Wetland
O   Open country
S   Scrub
B   Bare stony ground

Threats

1   Habitat loss or degradation
2   Hunting
3   Overfishing
4   Unknown
?   Threat considered likely, but evidence is only circumstantial

x   species occurs outside present or proposed protected areas

*   species requiring dense or moist conditions with plenty of undergrowth or trees covered in epiphytes

Black Bittern *Dupetor flavicollis*	V	W	1
Yellow Bittern *Ixobrychus sinensis*	V	W	1
Black Baza *Aviceda leuphotes*	E	*F	1
Pallas's Fish Eagle *Haliaeetus leucoryphus*	E	W	1,3
Lesser Fishing Eagle *Ichthyophaga nana*	E	FW	1,3
Grey-headed Fishing Eagle *Ichthyophaga ichthyaetus*	E	FW	1,3
Lesser Spotted Eagle *Aquila pomarina*	R	F	1,4
Tawny Eagle *Aquila rapax*	R	F	1,4
Rufous-bellied Eagle *Hieraaetus kienerii*	E	*F	1
Changeable Hawk-Eagle *Spizaetus cirrhatus*	V	*F	1
Red-necked Falcon *Falco chicquera*	E	F	4
Oriental Hobby *Falco severus*	R	F x	4
Laggar *Falco jugger*	R	O	4
Swamp Francolin *Francolinus gularis*	V	W	1
Black-breasted Quail *Coturnix coromandelica*	E	S x	2?,4
Blue-breasted Quail *Coturnix chinensis*	E	F	1,2?
Rufous-throated Hill Partridge *Arborophila rufogularis*	E	*F	1,2?
Satyr Tragopan *Tragopan satyra*	V	*F	1,2

Table 12 (contd)

Cheer Pheasant *Catreus wallichii*	I	F	2,4
Striped Buttonquail *Turnix sylvatica*	E	G	1,2?
Yellow-legged Buttonquail *Turnix tanki*	R	G	1,2?
Watercock *Gallicrex cinerea*	V	W	1
Bengal Florican *Houbaropsis bengalensis*	E	G	1,2?
Lesser Florican *Sypheotides indica*	E	G	1,2?
Indian Courser *Cursorius coromandelicus*	V	B x	1
Wood Snipe *Gallinago nemoricola*	I	*F	4
Barred Cuckoo-Dove *Macropygia unchall*	V	*F	1
Orange-breasted Green Pigeon *Treron bicincta*	R	*F	1
Pompadour Green Pigeon *Treron pompadora*	R	*F	1
Thick-billed Green Pigeon *Treron curvirostra*	V	*F	1
Pin-tailed Green Pigeon *Treron apicauda*	V	*F	1
Mountain Imperial Pigeon *Ducula badia*	E	*F	1
Vernal Hanging Parrot *Loriculus vernalis*	E	*F	1
Red-winged Crested Cuckoo *Clamator coromandus*	R	F	1
Asian Emerald Cuckoo *Chrysococcyx maculatus*	I	*F x	1
Banded Bay Cuckoo *Cacomantis sonneratii*	I	F	1
Grass Owl *Tyto capensis*	E	G	1
Mountain Scops Owl *Otus spilocephalus*	R	*F	1
Forest Eagle Owl *Bubo nipalensis*	E	*F	1
Dusky Eagle Owl *Bubo coromandus*	E	F	1
Brown Fish Owl *Ketupa zeylonensis*	V	*F	1,3
Tawny Fish Owl *Ketupa flavipes*	E	*F	1,3
Brown Wood Owl *Strix leptogrammica*	V	*F	1
White-rumped Needletail *Zoonavena sylvatica*	R	*F	1
White-vented Needletail *Hirundapus cochinchinensis*	I	?	4
Red-headed Trogon *Harpactes erythrocephalus*	E	*F	1
Ruddy Kingfisher *Halcyon coromanda*	E	*F x	1
Blyth's Kingfisher *Alcedo hercules*	I	*F x	1
Deep-blue Kingfisher *Alcedo meninting*	E	*F	1
Dollarbird *Eurystomus orientalis*	R	*F	1
Oriental Pied Hornbill *Anthracoceros coronatus*	V	*F	1,2
Great Pied Hornbill *Buceros bicornis*	E	*F	1,2
Blue-eared Barbet *Megalaima australis*	E	*F x	1
Orange-rumped Honeyguide *Indicator xanthonotus*	R	F	4
White-browed Piculet *Sasia ochracea*	R	*F	1
Pale-headed Woodpecker *Gecinulus grantia*	E	*F x	1
Bay Woodpecker *Blythipicus pyrrhotis*	V	*F	1
Great Slaty Woodpecker *Mulleripicus pulverulentus*	V	F	1
Long-tailed Broadbill *Psarisomus dalhousiae*	E	*F	1
Blue-naped Pitta *Pitta nipalensis*	E	*F x	1
Hooded Pitta *Pitta sordida*	R	*F	1
Indian Pitta *Pitta brachyura*	R	*F	1
Short-billed Minivet *Pericrocotus brevirostris*	R	F	1
Grey-chinned Minivet *Pericrocotus solaris*	E	F	1
Rosy Minivet *Pericrocotus roseus*	R	F	1
White-throated Bulbul *Criniger flaveolus*	V	*F	1
Asian Fairy Bluebird *Irena puella*	E	*F x	1

Table 12 (contd)

Gould's Shortwing *Brachypteryx stellata*	R	F	1
White-browed Shortwing *Brachypteryx montana*	R	*F	1
Lesser Shortwing *Brachypteryx leucophrys*	E	*F x	1
White-tailed Robin *Cinclidium leucurum*	R	*F	1
Purple Cochoa *Cochoa purpurea*	E	*F x	1
White-tailed Stonechat *Saxicola leucura*	R	G	1
Long-billed Thrush *Zoothera monticola*	R	*F	1
Dark-sided Thrush *Zoothera marginata*	I	*F	1
Slaty-bellied Tesia *Tesia olivea*	I	*F	1
Pale-footed Bush Warbler *Cettia pallidipes*	R	F	1
Yellow-bellied Bush Warbler *Cettia acanthizoides*	I	F	1
Bright-capped Cisticola *Cisticola exilis*	R	G	1
Large Grass Warbler *Graminicola bengalensis*	V	G	1
Bristled Grass Warbler *Chaetornis striatus*	I	G	1
Striated Marsh Warbler *Megalurus palustris*	V	G	1
Grey-cheeked Warbler *Seicercus poliogenys*	E	*F	1
Broad-billed Warbler *Abroscopus hodgsoni*	I	*F	1
Rufous-faced Warbler *Abroscopus albogularis*	E	*F x	1
Large Niltava *Niltava grandis*	V	*F	1
Pale Blue Flycatcher *Cyornis unicolor*	E	*F x	1
Hill Blue Flycatcher *Cyornis banyumas*	V	*F	1
Pygmy Blue Flycatcher *Muscicapella hodgsoni*	V	*F	1
Ferruginous Flycatcher *Muscicapa ferruginea*	R	*F	1
Sapphire Flycatcher *Ficedula sapphira*	V	*F x	1
Little Pied Flycatcher *Ficedula westermanni*	R	*F	1
White-gorgetted Flycatcher *Ficedula monileger*	V	*F	1
Abbott's Babbler *Trichastoma abbotti*	E	*F	1
Coral-billed Scimitar-Babbler *Pomatorhinus ferruginosus*	I	F	1
Slender-billed Scimitar-Babbler *Xiphirhynchus superciliaris*	V	*F	1
Spotted Wren-Babbler *Spelaeornis formosus*	I	*F	1
Tailed Wren-Babbler *Spelaeornis caudatus*	E	*F	1
Golden Babbler *Stachyris chrysaea*	E	*F	1
Great Parrotbill *Conostoma aemodium*	V	F	1
Brown Parrotbill *Paradoxornis unicolor*	V	F	1
Fulvous Parrotbill *Paradoxornis fulvifrons*	V	F	1
Black-throated Parrotbill *Paradoxornis nipalensis*	R	*F	1
Slender-billed Babbler *Turdoides longirostris*	V	G	1
Lesser Necklaced Laughing-thrush *Garrulax monileger*	V	*F	1
Greater Necklaced Laughing-thrush *Garrulax pectoralis*	V	*F	1
Rufous-chinned Laughing-thrush *Garrulax rufogularis*	V	*F	1
Grey-sided Laughing-thrush *Garrulax caerulatus*	E	*F	1
Rufous-necked Laughing-thrush *Garrulax ruficollis*	V	*F	1
Blue-winged Laughing-thrush *Garrulax squamatus*	E	*F	1
Scaly Laughing-thrush *Garrulax subunicolor*	V	*F	1
Silver-eared Mesia *Leiothrix argentauris*	E	*F x	1
Fire-tailed Myzornis *Myzornis pyrrhoura*	R	*F	1
Cutia *Cutia nipalensis*	E	*F	1
Black-headed Shrike-Babbler *Pteruthius rufiventer*	E	*F	1
White-hooded Shrike-Babbler *Gampsorhynchus rufulus*	E	*F x	1

Table 12 (contd)

Rusty-fronted Barwing *Actinodura egertoni*	E	*F	1
Golden-breasted Fulvetta *Alcippe chrysotis*	V	F	1
Chestnut-backed Sibia *Heterophasia annectans*	E	*F x	1
White-naped Yuhina *Yuhina bakeri*	E	*F x	1
Black-chinned Yuhina *Yuhina nigrimenta*	E	*F	1
Black-spotted Yellow Tit *Parus spilonotus*	E	*F x	1
Sultan Tit *Melanochlora sultanea*	E	*F x	1
Ruby-cheeked Sunbird *Anthreptes singalensis*	E	*F	1
Little Spiderhunter *Arachnothera longirostra*	E	*F	1
Yellow-vented Flowerpecker *Dicaeum chrysorrheum*	E	*F x	1
Yellow-bellied Flowerpecker *Dicaeum melanoxanthum*	R	F	1
Scarlet-backed Flowerpecker *Dicaeum cruentatum*	E	F x	1
Crow-billed Drongo *Dicrurus annectans*	V	*F	1
Black-breasted Weaver *Ploceus benghalensis*	R	G	1
Scarlet Finch *Haematospiza sipahi*	V	*F	1

some areas are being lost to cultivation, such as in the old river bed south of Kosi Barrage. The distribution of the Orange-rumped Honeyguide *Indicator xanthonotus* is linked with the nests of bees, as one of its main food items is beeswax (Cronin and Sherman 1976).

A total of 11 species (pheasants, quails and hornbills) are probably threatened to some degree by hunting, although habitat losses are thought to be much more important for most of them.

Overfishing was suggested as a threat to large fish-eating birds of prey in Chitwan (Thiollay 1978) and this could well be affecting them elsewhere in Nepal.

Several species are rare and considered to be at risk for unknown reasons. However, the widespread use of pesticides in the terai since the 1950s is likely to have reduced numbers of all the birds of prey which are now threatened.

Threatened species breeding only outside the currently protected areas or the two proposed protected areas (Annapurna Conservation Area and Barun valley) total 23, 18 percent of the total at risk.

The remaining 82 percent of threatened birds which breed within currently protected or proposed protected areas are still suffering from habitat losses or damage. An assessment of the disturbances, deficiencies and management problems of Nepal's protected areas system has been described by Green (1986).

Nepal may support internationally significant populations of as many as 31 of its threatened bird species and may be especially important for 20 of these. Three of them: Blue-naped Pitta *Pitta nipalensis*, Purple Cochoa *Cochoa purpurea* and Sapphire Flycatcher *Ficedula sapphira* are particularly at risk because they are not known to occur in present or proposed protected areas.

Orange-rumped Honeyguide *Indicator xanthonotus* (Dave Showler)

PROTECTED AREA COVERAGE

The great majority of Nepal's forest types are well represented in the protected area system and include all upper temperate, subalpine, alpine and most tropical forests types (Table 13).

Table 13: Protected area coverage of Nepal's main forest types.

Column 1 Conservation achievements – well protected
Column 2 Unprotected or inadequately represented

Forest type	1	2
Tropical: East tropical evergreen component		+
Tropical: Central	+	
Tropical: Far west	+	
Subtropical: Far eastern broadleaves		
Schima-Castanopsis and subtropical evergreen		+
Subtropical: *Schima-Castanopsis* further west		+
Subtropical: West Nepal	+	
Lower temperate: Mixed broadleaves and *Quercus lamellosa* in far east		+
Lower temperate: *Quercus lamellosa* and mixed broadleaves further west	+	
Lower temperate: *Quercus lanata* in centre and east		+
Lower temperate: West Nepal	+	
Upper temperate: Far eastern broadleaves	+	
Upper temperate: Central broadleaves	+	
Upper temperate: Western broadleaves	+	
Upper temperate: Western conifers	+	
Subalpine: *Abies spectabilis*	+	
Subalpine: *Betula utilis*	+	
Subalpine: *Rhododendron* spp.	+	
Subalpine: *Juniperus* spp.	+	
Alpine	+	

An outstanding gap is that of the subtropical broadleaved forests made up of evergreen forests and *Schima/Castanopsis* with associated riverine forests. There is a special case for protecting *Schima/Castanopsis* as this once covered much of central and eastern Nepal. The only protected or proposed protected forests of these types are small areas of *Schima wallichii/Castanopsis tribuloides* in the Barun valley and of *Schima wallichii/Castanopsis indica* in the Annapurna Conservation Area and Langtang National Park. The forests in the latter two areas are now much degraded (Kenting 1986). Two types of subtropical broadleaved forest are worthy of protection, one in the far east in the Mai valley area and the other further west (see p.47).

Tropical evergreen forest is not included in the protected area system apart from a tiny area in the Royal Chitwan National Park.

While the species-rich *Quercus lamellosa* forests are quite well represented, some other lower temperate forest types are inadequately protected or unprotected. Those in the Mai valley area of far east Nepal warrant protection of a representative area. *Quercus lanata* forests in central and eastern Nepal are poorly covered by parks and reserves. However, the small remnants described as in reasonable condition by Stainton (1972) may not be worthy of protection, especially as they may well have deteriorated since the early 1970s. The only protected western lower temperate forests occur in Khaptad National Park (Dobremez 1974-1985), and they are of high quality.

There are just 29 breeding bird species in Nepal which are currently recorded only outside the protected and proposed protected areas in the nesting season (Table 14). Four are non-forest species: the threatened Indian Courser *Cursorius coromandelicus* (described on p.61), the Indian Skimmer *Rynchops albicollis*, Barn

Table 14: Breeding birds recorded only outside Nepal's protected and proposed protected areas (all species).

Oriental Hobby *Falco severus*
Black-breasted Quail *Coturnix coromandelica*
Indian Courser *Cursorius coromandelicus*
Indian Skimmer *Rynchops albicollis*
Asian Emerald Cuckoo *Chrysococcyx maculatus*
Barn Owl *Tyto alba*
Asian Palm Swift *Cypsiurus balasiensis*
Ruddy Kingfisher *Halcyon coromanda*
Blyth's Kingfisher *Alcedo hercules*
Blue-eared Barbet *Megalaima australis*
Pale-headed Woodpecker *Gecinulus grantia*
Blue-naped Pitta *Pitta nipalensis*
Asian Fairy Bluebird *Irena puella*
Lesser Shortwing *Brachypteryx leucophrys*
Purple Cochoa *Cochoa purpurea*
Hill Prinia *Prinia atrogularis*
Rufous-faced Warbler *Abroscopus albogularis*
Pale Blue Flycatcher *Cyornis unicolor*
Sapphire Flycatcher *Ficedula sapphira*
Common Babbler *Turdoides caudatus*
Large Grey Babbler *Turdoides malcolmi*
Silver-eared Mesia *Leiothrix argentauris*
White-hooded Babbler *Gampsorhynchus rufulus*
Chestnut-backed Sibia *Heterophasia annectans*
White-naped Yuhina *Yuhina bakeri*
Black-spotted Yellow Tit *Parus spilonotus*
Sultan Tit *Melanochlora sultanea*
Yellow-vented Flowerpecker *Dicaeum chrysorrheum*
Scarlet-backed Flowerpecker *Dicaeum cruentatum*

Owl *Tyto alba* and Asian Palm Swift *Cypsiurus balasiensis*.   Forest species
number 25 and four of these occur in scrub.   All are tropical, subtropical or
lower temperate birds apart from Sapphire Flycatcher *Ficedula sapphira* which
probably also breeds in upper temperate forests.   A total of 23 of them are
thought to be at risk.   It is possible that some of the subtropical and lower
temperate species may occur in the Barun valley which is particularly
under-recorded.   Lesser Shortwing *Brachypteryx leucophrys* and Sapphire
Flycatcher *Ficedula sapphira* have been located in the valley of the Arun, of
which the Barun forms a tributary.   Nine species associated with tropical
evergreen forest occur outside parks and reserves (Table 11) and five species are
rare at Chitwan and are more common in the more extensive tropical evergreen
forests further east.

Nine subtropical forest species only occur outside protected or proposed
protected areas.   Four of them: Asian Emerald Cuckoo *Chrysococcyx maculatus*,
Blue-naped Pitta *Pitta nipalensis*, Pale Blue Flycatcher *Cyornis unicolor* and
White-naped Yuhina *Yuhina bakeri* are restricted to these forests.   The other five
species which are unrepresented occur in the subtropical and tropical or
subtropical and lower temperate zones.   These are Purple Cochoa *Cochoa
purpurea*, Hill Prinia *Prinia atrogularis*, Chestnut-backed Sibia *Heterophasia
annectans*, Silver-eared Mesia *Leiothrix argentauris* and Sultan Tit *Melanochlora
sultanea*.   The latter two species prefer the subtropical zone.

There are six lower temperate species which have only been found outside
protected areas and all occur in the east: Lesser Shortwing *Brachypteryx
leucophrys*, Purple Cochoa *Cochoa purpurea*, Hill Prinia *Prinia atrogularis*,
Sapphire Flycatcher *Ficedula sapphira*, Chestnut-backed Sibia *Heterophasia
annectans* and Black-spotted Yellow Tit *Parus spilonotus*.

With the exception of the Hill Prinia (a species of grassland and scrub) all the
above-mentioned species of the tropical, subtropical and lower temperate forests
require wet conditions and all are at risk.   Hill Prinia *Prinia atrogularis*,
Chestnut-backed Sibia *Heterophasia annectans*, White-naped Yuhina *Yuhina bakeri*
and Black-spotted Yellow Tit *Parvus spilonotus* are only known from the
subtropical and/or lower temperate forests in the Mai valley and Hans Pokhari
Danda in the far east.   Blue-naped Pitta *Pitta nipalensis*, Purple Cochoa *Cochoa
purpurea* and Sapphire Flycatcher *Ficedula sapphira* may have significant
breeding populations in Nepal.

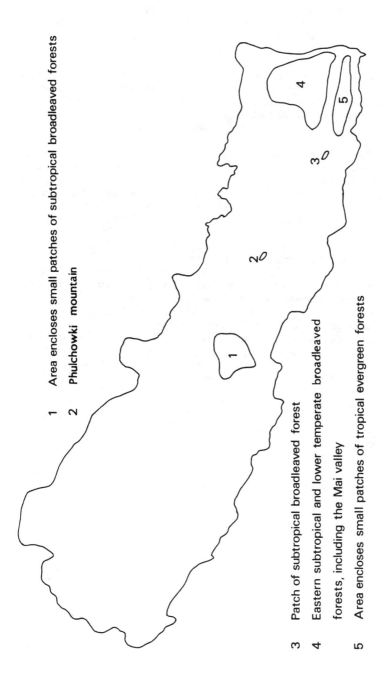

1   Area encloses small patches of subtropical broadleaved forests

2   **Phulchowki mountain**

3   Patch of subtropical broadleaved forest

4   Eastern subtropical and lower temperate broadleaved forests, including the Mai valley

5   Area encloses small patches of tropical evergreen forests

Figure 9: Nepal's important unprotected forest areas.

SOME IMPORTANT
UNPROTECTED FORESTS

Some forests which are particularly rich in bird species and therefore of national importance were identified by Inskipp and Inskipp (1986): the southern slopes of Annapurna, Tansen area, south-east corner of the Kathmandu valley (Phulchowki mountain), upper and lower Arun valley and the Mai valley (Figure 9). These areas have exceptionally high rainfall as a result of local topographical features and support luxuriant forests. They are also likely to support a high species diversity.

The southern slopes of Annapurna and the most important part of the upper Arun watershed lie within the proposed protected areas − the Barun valley and Annapurna Conservation Area. The LRMP land utilisation maps showed that in 1978/79 the Tansen and Bagmati valley forests had either been converted to agriculture or badly degraded (reduced to less than 40 percent tree crown cover). However, significant patches of *Schima wallichii/Castanopsis tribuloides* with some *C. indica* remained in the lower Arun valley between 26°45'N and 27°15'N, and 87°E and 87°15'E (LRMP map numbers N1 and M4). These have been almost unstudied ornithologically. The two most important unprotected species-rich forests are those in the Mai valley and on Phulchowki mountain.

MAI VALLEY

Site description
The Mai valley (70-3,050 m) lies in far eastern Nepal in the tropical, subtropical and both temperate zones.

Importance for birds
The valley's tropical zone includes examples of unprotected tropical evergreen forests (which extends west to the Kosi river). The great importance of this forest type and the need for a representative protected area are described on pages 47 and 68. The following account is therefore limited to forests at higher altitudes. A bird species inventory annotated with status is given on p.149, although more fieldwork is needed to compile a comprehensive list. The large total of 297 species has been recorded so far, including eight found only by Stevens (1923, 1924a,b, 1925a,b,c), all of which could still occur. These species are not counted in the following analysis. Breeding species total 232. Half of the species (63) for which Nepal may hold internationally significant populations breed there and the country may be especially important for 11 of these. The forests support 34 or 35 breeding species considered at risk in Nepal including 14 in the endangered category. There are as many as seven or eight of the 29 breeding species currently only recorded outside protected or proposed protected areas: Lesser Shortwing *Brachypteryx leucophrys*, Purple Cochoa *Cochoa*

purpurea, Hill Prinia *Prinia atrogularis*, Sapphire Flycatcher *Ficedula sapphira*, Silver-eared Mesia *Leiothrix argentauris*, *Chestnut-backed Sibia *Heterophasia annectans*, *Black-spotted Yellow Tit *Parus spilonotus* and possibly Sultan Tit *Melanochlora sultanea* (only recorded once, Walinder and Sandgren 1982). Two of these species (marked *) have only been recorded from these forests and all are considered endangered. The forests are also important for winter visitors which number 55 and include 12 which may have significant breeding populations in Nepal.

Threats
These have been little documented, but were being adversely affected by removal of food and foliage and grazing of livestock in 1981(pers. obs.) and 1986 (S. Gawn *in litt.* 1987). Logging was severely depleting the tropical and subtropical forests in 1988 (van Riessen pers. comm.).

Future conservation
The protection of the valley's forests or a representative area is recommended.

PHULCHOWKI MOUNTAIN

Site description
Phulchowki mountain (1,525-2,760 m), the highest peak on the rim of the Kathmandu valley lies 16km south-east of Kathmandu and covers an area of about 50 sq km. Its slopes receive exceptionally high rainfall and support luxuriant forests comprising subtropical broadleaved *Schima wallichii/Castanopsis indica* forests on the lower slopes, *Quercus lamellosa* and *Q. lanata* higher up and *Q. semecarpifolia* around the summit.

Importance for birds
A bird species list annotated with status is given on p.155. As Phulchowki's forests are a popular destination for birdwatchers, they have been well-recorded throughout the year. The high total of 256 bird species has been found including 155 species that breed or probably breed, 92 of which breed in subtropical forests. As many as 90 percent of the total recorded are dependent on forests. Among the breeding birds there are 35 which may have internationally significant populations in Nepal and 17 considered to be at risk in the country with six in the endangered category: Rufous-throated Hill Partridge *Arborophila rufogularis*, Blue-naped Pitta *Pitta nipalensis*, Grey-chinned Minivet *Pericrocotus solaris*, Grey-sided Laughing-thrush *Garrulax caerulatus*, Blue-winged Laughing-thrush *Garrulax squamatus* and Cutia *Cutia nipalensis*.

The forests are also of considerable importance for wintering birds and passage migrants or visiting species which number 51 and 35 respectively, discounting vagrants.

There are 23 winter visitors which may have significant breeding populations in Nepal, including a regular flock of the Spot-winged Grosbeak *Mycerobas melanozanthos*, a species for which Nepal may be especially important. Three threatened species winter on the mountain: Long-billed Thrush *Zoothera monticola*, Yellow-bellied Flowerpecker *Dicaeum melanoxanthum* and Scarlet Finch *Haematospiza sipahi*.

Importance for other wildlife

Phulchowki's forests are internationally famous for the variety of their other wildlife too. Martens (1979) states that 'Numerous animal species, especially insects and *Arachnida* hitherto unknown to science have been discovered here in recent years'. The forests are also of great value for their flora. Recent research has revealed that a few endemic plants grow there and it is the only Nepalese locality for some other plant species (Shrestha pers. comm.). Ghimre (1984-1985) advocates their complete protection on the basis of their botanical importance.

Threats

Severe and increasing threats face Phulchowki's forests.

Quarries

The subtropical forests are especially threatened by quarries on the lower slopes. Since about 1975 they have been extensively quarried and only bare rock remains over large sections. Many workers' homes and a factory have been erected below the quarry on land which was forest only a few years ago.

Firewood

Phulchowki is now the main source of firewood in the valley. Local wood-cutting parties daily remove large quantities either for their own use or for sale in Kathmandu. *Quercus semecarpifolia*, *Q. incana*, *Schima wallichii*, *Rhododendron arboreum*, *Castanopsis indica*, *Alnus nepalensis*, *Myrica esculenta* and *Lyonia ovalifolia* are the main tree species taken (Khadka *et al.* 1984-1985).

Charcoal

Large quantities of wood, almost entirely *Lyonia ovalifolia* are used to make charcoal. An aerial photograph taken in 1978 showed a total area of 0.42 sq km devastated by these activities (Khadka *et al.* 1984-1985).

Road

The road which runs from the mountain's base to its summit is now surfaced for much of its length and allows vehicles to easily remove timber from the upper as well as lower slopes. In autumn 1986 and spring 1988 there was evidence of many mature trees being transported in this way (pers. obs.).

Animal fodder

Enormous quantities of foliage are collected for this purpose. The trees *Castanopsis indica*, *Quercus semecarpifolia* and *Q. incana* are favoured.

These threats have been well described elsewhere (e.g. by Dixit 1986; Ghimre 1984-1985; Khadka *et al.* 1984-1985; Martens 1983).

Future conservation

Despite these threats there is still time to save and conserve Phulchowki's forests and their wildlife. Almost all previously recorded breeding and wintering bird species still occurred up to at least 1986 (discounting vagrants). The exceptions are Wood Snipe *Gallinago nemoricola*, Red-headed Trogon *Harpactes erythrocephalus* and Blue-naped Pitta *Pitta nipalensis*. Wood Snipe was collected last century by Hodgson (1829) and Red-headed Trogon was found breeding by

Proud (1955), but there are no other records of either species from the mountain. Blue-naped Pitta was last seen in 1983 (Holmström 1983), but could be overlooked as it is secretive and rare. Rufous-throated Hill Partridge *Arborophila rufogularis* was considered to be extinct there by 1982 (Fleming, pers. comm.), but two reliable reports have been received since (Parr 1982; Heath 1986).

However, if Phulchowki's forests and their rich variety of flora and fauna are to continue to survive, their protection is urgently needed and the quarries must be closed. At the present rate of destruction the stage will be reached in the foreseeable future where many species will disappear. Lying only 40 minutes drive from the busy centre of Kathmandu, Phulchowki could, with protection, become a valuable retreat for both Nepalese and tourists. Tourism could provide valuable revenue for both the Department of National Parks and Wildlife Conservation and the local people who could act as guides and forest guards. Walking up the road to the summit is an easy way of seeing a great variety of Nepal's birds, many of which are otherwise difficult to observe, and this method results in minimal disturbance.

IDENTIFICATION OF OTHER UNPROTECTED FORESTS IMPORTANT FOR BIRDS AND OTHER WILDLIFE

Forest areas existing in 1978/79 were located using a combination of Dobremez's vegetational maps (1974-1985) and the land utilisation maps (scale 1:50,000 produced by the LRMP (Kenting 1986). The LRMP maps include the extent of forest areas, dominant tree species or a description of forest type and crown density (expressed as a percentage of the area covered by tree crowns).

The forest types unrepresented or poorly represented in the protected area system (tropical evergreen, far eastern subtropical and lower temperate broadleaved forests, and subtropical broadleaved forests further west) were identified on the LRMP maps in the following way. Tropical evergreen forests lie within forests classified as sal or tropical mixed hardwoods with 75 percent or more hardwood tree species. Subtropical broadleaved forests are classified within tropical mixed hardwoods and lower temperate forests as deciduous mixed broadleaves each with more than 75 percent hardwoods.

Forests considered worthy of protection had crown densities classified as: Category 3 – 40-70 percent of the area covered by tree crowns; or Category 4 – more than 70 percent of the area covered by tree crowns.

All the LRMP land utilisation maps were examined in the appropriate climatic zone to locate suitable forest areas fitting the above descriptions. The tropical zone between the Kosi and Mechi rivers in the south-east (where almost all of this forest type grows) was searched for areas which may contain tropical evergreen forest, the subtropical zone throughout Nepal was examined for subtropical broadleaved forests and the lower temperate zone east of 87 30'E was searched for broadleaved forests. Dobremez's vegetational maps were then used to identify the forest types within these areas.

Only relatively small patches remained in 1978/79. The areas requiring surveys are described in the recommendations section that follows.

RECOMMENDATIONS

1. Protection of a subtropical broadleaved forest (i.e. *Schima-Castanopsis*), typical of those which once occurred throughout much of central and eastern Nepal, is highly recommended. A very good example would on Phulchowki mountain in the Kathmandu valley.

2. Survey work is needed in the remaining forests of the types unrepresented in the protected area system to enable identification of new areas suitable for protection. The location of these forests which were in reasonable condition (i.e. with more than 40 percent crown cover) in 1978/79 and the respective LRMP map numbers are given below. Copies of the LRMP maps marking these forests are available from the author, c/o ICBP. Fieldwork is required to find out which of them still exist, to assess their quality and to survey their birds and other wildlife groups so that representative areas can be chosen.
    Unprotected forest types and locations of remaining areas in reasonable condition in 1978/79:

Other *Schima-Castanopsis* forests (in case Phulchowki cannot be protected):

Schima wallichii/Castanopsis indica mainly in west-central Nepal near Pokhara between 27°45'N and 28°30'N and 84°30'E and 83°45'E (LRMP map numbers 72A1, 72A5, 71D3, 71D4, 71D7, 71D8, 62P15, 62P16) also a significant area between 26°45'N and 27°N and 86°45'E and 86°30'E (LRMP map number 72J9) and two small patches in the Kathmandu valley (LRMP map number E6).

Smaller areas of *Schima wallichii/Castanopsis tribuloides* with some *C. indica* in the east between 26°45'N and 27°30'N and 87°E and 88°15'E (LRMP map numbers 72M4, 72M7, 72M8, 72M11, 72M16, 72N1, 72N5, 72N9, 72N13, 78B1). Forests between 26°45'N and 27°15'N and 87°E and 87°15'E (LRMP map numbers 72N1, 72M4) are likely to be species-rich and are therefore particularly worthy of investigation.

Tropical evergreen forest. Within sal forest in shady ravines and north-facing gulleys between the Kosi and Mechi rivers in south-east Nepal between 26°15'N and 27°N and 87°E and 88°15'E (LRMP map numbers 72N1, 72N5, 72N6, 72N10, 72N13, 72N14, 78B1, 78B2, 78B3).

Subtropical and lower temperate forest in far eastern Nepal:

Subtropical forests between 26°45'N and 27°15'N and 87°45'E and 88°15'E (LRMP map numbers 72M16, 72N13, 78B1).

Lower temperate forests between 27°N and 27°45'N and 87°30'E and 88°15'E (LRMP map numbers 72M10, 72M11, 72M12, 72M14, 72M15, 72M16, 78A3, 78A4).

3. As Nepalese forests face severe threats and only relatively small patches of the unprotected forest types remain, all the necessary survey work and designation of new protected areas are urgently needed.

4. More fieldwork is required in western Nepal forests (west of the Kali-Gandaki river) as they are under-recorded.

5. More bird surveys are needed in all the under-recorded protected areas: Khaptad, Rara and Shey-Phoksundo National Parks, Royal Sukla Phanta, Royal Bardia, Shivapuri Watershed and Kosi Tappu Wildlife Reserves and Dhorpatan Hunting Reserve and especially in the species-rich Barun valley — a proposed protected area. A good knowledge of the occurrence and status of bird species will be useful for the better management of these areas. It will also enable the identification of bird species still requiring habitat protection to be re-assessed.

6. A study to identify the forests important for birds outside the breeding season is needed.

CONCLUSION

Nepal has an extensive protected area system which covers 7.4 percent of the country. When the proposed protected areas – the Annapurna Conservation Area and Barun valley extension to Sagarmatha National Park – are designated this figure will increase to over 10 percent. While the majority of Nepal's forest types are well represented, there are a few important omissions. The most significant is the subtropical broadleaved forest, *Schima-Castanopsis*, which is especially valuable as part of the country's natural heritage as it once covered much of central and eastern Nepal. Protection of examples of tropical evergreen and far eastern subtropical and lower temperate forests are also needed.

There are just 29 breeding birds (all habitats) which are currently recorded only outside the existing and proposed protected areas in the nesting season. However, a large number of breeding birds – 22 percent of the total – are considered to be at risk and 84 percent of these are dependent on forests. Forest losses and deterioration are by far the greatest threats to Nepal's birds. Once Nepal was extensively forested, but by 1979 only 42.8 percent of the country was 'forest land' (covered in trees and shrubs). A large percentage of this was covered in forests of low density and only 28.1 percent of the country had more than 40 percent tree cover. Forest resources continue to decline, chiefly because they can no longer meet requirements of the population for their basic needs. Conservation of the country's forests is therefore vital for the future of its people as well as its birds.

There has recently been a great expansion in afforestation, but the overall impact has been very small. As far as most birds are concerned the new plantations are preferable to areas devoid of forest. However, plantations, even of native broadleaves, can never replace the richness and variety of natural forests which may have taken hundreds of years to develop. Many people in Nepal now believe that the most important aspect of forestry in the country is the improved management of the large existing areas of low density forest. This policy has enormous potential and would be much more valuable to forest birds than planting more trees.

Blue-fronted Redstart *Phoenicurus frontalis* (Dave Showler)

BIRD CHECKLISTS

Key to Checklists

1   Common

2   Fairly common

3   Occasional

4   Uncommon

5   Rare

b   Proved breeding

m   Passage migrant

r   Resident

s   Summer visitor

w   Winter visitor

v   Vagrant

?   Status or abundance uncertain

+   Species for which Nepal may hold internationally significant breeding populations

*   Species for which Nepal may hold internationally significant breeding populations and for which the country may be especially important

E   Endangered

V   Vulnerable

R   Rare

I   Indeterminate

ROYAL CHITWAN NATIONAL PARK
based on Gurung's checklist (1983) with recent additions

Little Grebe *Tachybaptus ruficollis*		w 5
Great Crested Grebe *Podiceps cristatus*		w 5
Great Cormorant *Phalacrocorax carbo*		w 1
Little Cormorant *Phalacrocorax niger*		w 4
Darter *Anhinga melanogaster*		r 1
Eurasian Bittern *Botaurus stellaris*		v
V	Yellow Bittern *Ixobrychus sinensis*	s 4
Cinnamon Bittern *Ixobrychus cinnamomeus*		r s 3
Black Bittern *Dupetor flavicollis*		v
Black-crowned Night Heron *Nycticorax nycticorax*		r s 1
Green-backed Heron *Butorides striatus*		r s 2
Indian Pond Heron *Ardeola grayii*		r 1
Cattle Egret *Bubulcus ibis*		r 2
Little Egret *Egretta garzetta*		r 1
Intermediate Egret *Egretta intermedia*		r 2
Great Egret *Egretta alba*		r 2
Grey Heron *Ardea cinerea*		r 5 w 1
Purple Heron *Ardea purpurea*		r 5 w 2
Painted Stork *Mycteria leucocephala*		s 5
Asian Openbill Stork *Anastomus oscitans*		r 1
Black Stork *Ciconia nigra*		w 2
Woolly-necked Stork *Ciconia episcopus*		r 3
White Stork *Ciconia ciconia*		v
Black-necked Stork *Ephippiorrhynchus asiaticus*		w 2
Greater Adjutant Stork *Leptoptilos dubius*		v
Lesser Adjutant Stork *Leptoptilos javanicus*		r 3
Red-naped Ibis *Pseudibis papillosa*		r 1
Lesser Whistling Duck *Dendrocygna javanica*		r 1
Tundra Swan *Cygnus columbianus*		v
Bean Goose *Anser fabalis*		v
Greylag Goose *Anser anser*		w m 4
Bar-headed Goose *Anser indicus*		w m 3
Ruddy Shelduck *Tadorna ferruginea*		w 1
Common Shelduck *Tadorna tadorna*		w 5
Cotton Pygmy Goose *Nettapus coromandelianus*		w 4
Eurasian Wigeon *Anas penelope*		m 4
Falcated Duck *Anas falcata*		w 5
Gadwall *Anas strepera*		w 4
Common Teal *Anas crecca*		w 2
Mallard *Anas platyrhynchos*		w 4
Spotbill *Anas poecilorhyncha*		w 4
Northern Pintail *Anas acuta*		w 3
Garganey *Anas querquedula*		m 5
Northern Shoveler *Anas clypeata*		w? m? 4
Red-crested Pochard *Netta rufina*		w m 4
Common Pochard *Aythya ferina*		w m 4
Ferruginous Duck *Aythya nyroca*		v
Tufted Duck *Aythya fuligula*		m 4

	Common Goldeneye *Bucephala clangula*	w 5
	Smew *Mergus albellus*	v
	Goosander *Mergus merganser*	w 2
E	Black Baza *Aviceda leuphotes*	s 5
	Crested Honey Buzzard *Pernis ptilorhyncus*	r 2
	Black-shouldered Kite *Elanus caeruleus*	r 3
	Black Kite *Milvus migrans*	w 1
	Brahminy Kite *Haliastur indus*	w 5
E	Pallas's Fish Eagle *Haliaeetus leucoryphus*	w 5
	White-tailed Eagle *Haliaeetus albicilla*	w 5
E	Lesser Fishing Eagle *Ichthyophaga nana*	r 5
E	Grey-headed Fishing Eagle *Ichthyophaga ichthyaetus*	r 5
	Egyptian Vulture *Neophron percnopterus*	w 3
	Oriental White-backed Vulture *Gyps bengalensis*	r 1
	Long-billed Vulture *Gyps indicus*	r 2
	Eurasian Griffon Vulture *Gyps fulvus*	r 2
	Red-headed Vulture *Sarcogyps calvus*	w 4
	Eurasian Black Vulture *Aegypius monachus*	w 4
	Short-toed Eagle *Circaetus gallicus*	w? m? 4
	Crested Serpent Eagle *Spilornis cheela*	r 1
	Eurasian Marsh Harrier *Circus aeruginosus*	w 2
	Hen Harrier *Circus cyaneus*	w 2
	Pallid Harrier *Circus macrourus*	w 5
	Montagu's Harrier *Circus pygargus*	m 5
	Pied Harrier *Circus melanoleucus*	w 3
	Northern Goshawk *Accipiter gentilis*	w 5
	Besra *Accipiter virgatus*	w? 5
	Northern Sparrowhawk *Accipiter nisus*	w? m? 5
	Crested Goshawk *Accipiter trivirgatus*	r 4
	Shikra *Accipiter badius*	r 2
	White-eyed Buzzard *Butastur teesa*	r 3
	Common Buzzard *Buteo buteo*	w m 5
	Long-legged Buzzard *Buteo rufinus*	w m 5
	Black Eagle *Ictinaetus malayensis*	w 5
R	Lesser Spotted Eagle *Aquila pomarina*	r 5
	Greater Spotted Eagle *Aquila clanga*	w 4
	Steppe Eagle *Aquila rapax nipalensis*	w m 2
R	Tawny Eagle *Aquila rapax vindhiana*	r 5
	Imperial Eagle *Aquila heliaca*	w? m? 5
	Booted Eagle *Hieraaetus pennatus*	w m 4
E	Rufous-bellied Eagle *Hieraaetus kienerii*	r? 5
V	Changeable Hawk-Eagle *Spizaetus cirrhatus*	r 4
	Mountain Hawk-Eagle *Spizaetus nipalensis*	w 4
	Osprey *Pandion haliaetus*	r w 2
	Red-thighed Falconet *Microhierax caerulescens*	r 4
	Lesser Kestrel *Falco naumanni*	m 5
	Common Kestrel *Falco tinnunculus*	w 4
E	Red-necked Falcon *Falco chicquera*	w? 5
	Amur Falcon *Falco amurensis*	m 5
	Eurasian Hobby *Falco subbuteo*	w m 5
R	Oriental Hobby *Falco severus*	w? m? 5
R	Laggar *Falco jugger*	r? 5

	Peregrine *Falco peregrinus*	w 5
	Black Francolin *Francolinus francolinus*	r 2
	Common Quail *Coturnix coturnix*	r? 2
E	Blue-breasted Quail *Coturnix chinensis*	r 5
	Red Junglefowl *Gallus gallus*	r 1
	Kalij Pheasant *Lophura leucomelana*	r 3
	Blue Peafowl *Pavo cristatus*	r 1
E	Striped Buttonquail *Turnix sylvatica*	r 5
R	Yellow-legged Buttonquail *Turnix tanki*	r 5
	Barred Buttonquail *Turnix suscitator*	r 3
	Baillon's Crake *Porzana pusilla*	w 5
	Ruddy-breasted Crake *Porzana fusca*	r 2
	Brown Crake *Amaurornis akool*	r 1
	White-breasted Waterhen *Amaurornis phoenicurus*	r 1
	Common Moorhen *Gallinula chloropus*	r w 1
	Purple Gallinule *Porphyrio porphyrio*	w 2
	Common Coot *Fulica atra*	w 4
	Common Crane *Grus grus*	w m 2
	Sarus Crane *Grus antigone*	v
	Demoiselle Crane *Anthropoides virgo*	m 2
E*	Bengal Florican *Houbaropsis bengalensis*	r 5
E	Lesser Florican *Sypheotides indica*	r? s? 5
	Pheasant-tailed Jacana *Hydrophasianus chirurgus*	s 5
	Bronze-winged Jacana *Metopidius indicus*	r 2
	Painted Snipe *Rostratula benghalensis*	r 4
	Black-winged Stilt *Himantopus himantopus*	w m 5
	Pied Avocet *Recurvirostra avosetta*	v
	Northern Stone-curlew *Burhinus oedicnemus*	r 2
	Great Stone-Plover *Esacus recurvirostris*	r 3
	Oriental Pratincole *Glareola maldivarum*	v
	Little Pratincole *Glareola lactea*	r 1
	Little Ringed Plover *Charadrius dubius*	r w 1
	Kentish Plover *Charadrius alexandrinus*	w 3
	Lesser Sand Plover *Charadrius mongolus*	v
	Pacific Golden Plover *Pluvialis fulva*	m 4
	River Plover *Hoplopterus duvaucelii*	r 1
	Yellow-wattled Plover *Hoplopterus malabaricus*	w 5
	Grey-headed Plover *Hoplopterus cinereus*	v
	Red-wattled Plover *Hoplopterus indicus*	r 1
	Northern Lapwing *Vanellus vanellus*	w 5
	Little Stint *Calidris minuta*	w m 4
	Temminck's Stint *Calidris temminckii*	w m 1
	Curlew Sandpiper *Calidris ferruginea*	w m 5
	Dunlin *Calidris alpina*	w m 4
	Ruff *Philomachus pugnax*	w m 5
	Jack Snipe *Lymnocryptes minimus*	w? m? 5
	Common Snipe *Gallinago gallinago*	w m 2
	Pintail Snipe *Gallinago stenura*	w m 3
	Eurasian Woodcock *Scolopax rusticola*	w 5
	Whimbrel *Numenius phaeopus*	m 5
	Eurasian Curlew *Numenius arquata*	m 5
	Spotted Redshank *Tringa erythropus*	w m 3

	Common Redshank *Tringa totanus*	w m 3
	Marsh Sandpiper *Tringa stagnatilis*	w m 4
	Common Greenshank *Tringa nebularia*	w m 1
	Green Sandpiper *Tringa ochropus*	w m 1
	Wood Sandpiper *Tringa glareola*	w m 3
	Common Sandpiper *Actitis hypoleucos*	w m 1
	Great Black-headed Gull *Larus ichthyaetus*	w m 2
	Common Black-headed Gull *Larus ridibundus*	w m 5
	Brown-headed Gull *Larus brunnicephalus*	w m 4
	Gull-billed Tern *Gelochelidon nilotica*	v
	Caspian Tern *Sterna caspia*	w m 5
	River Tern *Sterna aurantia*	r 2
	Common Tern *Sterna hirundo*	m 5
	Black-bellied Tern *Sterna acuticauda*	r s? 1
	Little Tern *Sterna albifrons*	s 2
	Whiskered Tern *Chlidonias hybridus*	v
	White-winged Black Tern *Chlidonias leucopterus*	v
	Rock Pigeon *Columba livia*	r 2
	Eurasian Collared Dove *Streptopelia decaocto*	r 2
	Red Turtle Dove *Streptopelia tranquebarica*	r 2
	Oriental Turtle Dove *Streptopelia orientalis*	w 2
	Laughing Dove *Streptopelia senegalensis*	m 5
	Spotted Dove *Streptopelia chinensis*	r 1
	Emerald Dove *Chalcophaps indica*	r 1
R	Orange-breasted Green Pigeon *Treron bicincta*	r 1
R	Pompadour Green Pigeon *Treron pompadora*	r 2
V	Thick-billed Green Pigeon *Treron curvirostra*	r 5
	Yellow-footed Green Pigeon *Treron phoenicoptera*	r 1
V	Pin-tailed Green Pigeon *Treron apicauda*	r 5
E	Mountain Imperial Pigeon *Ducula badia*	r 5
E	Vernal Hanging Parrot *Loriculus vernalis*	r 5
	Alexandrine Parakeet *Psittacula eupatria*	r 1
	Ring-necked Parakeet *Psittacula krameri*	r 1
+	Slaty-headed Parakeet *Psittacula himalayana*	w 5
	Blossom-headed Parakeet *Psittacula cyanocephala*	r 1
	Moustached Parakeet *Psittacula alexandri*	r 2
	Pied Crested Cuckoo *Clamator jacobinus*	s 4
	Red-winged Crested Cuckoo *Clamator coromandus*	s 2
	Common Hawk-Cuckoo *Hierococcyx varius*	r 1
	Large Hawk-Cuckoo *Hierococcyx sparverioides*	v
I	Asian Emerald Cuckoo *Chrysococcyx maculatus*	v
	Grey-bellied Plaintive Cuckoo *Cacomantis passerinus*	s 4
	Rufous-bellied Plaintive Cuckoo *Cacomantis merulinus*	v
I	Banded Bay Cuckoo *Cacomantis sonneratii*	r? s? 4
	Indian Cuckoo *Cuculus micropterus*	s 1
	Common Cuckoo *Cuculus canorus*	s m 3
	Drongo Cuckoo *Surniculus lugubris*	s 1
	Common Koel *Eudynamys scolopacea*	s? 4
	Green-billed Malkoha *Phaenicophaeus tristis*	r 2
	Sirkeer Malkoha *Phaenicophaeus leschenaultii*	r? 5
	Greater Coucal *Centropus sinensis*	r 1
	Lesser Coucal *Centropus bengalensis*	r s 2

E	Grass Owl *Tyto capensis*	r 5
	Indian Scops Owl *Otus bakkamoena*	r
E	Forest Eagle Owl *Bubo nipalensis*	r 5
E	Dusky Eagle Owl *Bubo coromandus*	r 5
V	Brown Fish Owl *Ketupa zeylonensis*	r 4
E	Tawny Fish Owl *Ketupa flavipes*	r 5
	Jungle Owlet *Glaucidium radiatum*	r 1
	Asian Barred Owlet *Glaucidium cuculoides*	r 2
	Brown Hawk Owl *Ninox scutulata*	r 2
	Spotted Little Owl *Athene brama*	r 3
	Short-eared Owl *Asio flammeus*	w? m? 5
	Savanna Nightjar *Caprimulgus affinis*	r? s? 2
	Indian Nightjar *Caprimulgus asiaticus*	r? s? 5
	Large-tailed Nightjar *Caprimulgus macrurus*	r? s? 2
	Jungle Nightjar *Caprimulgus indicus*	m? 2
	Himalayan Swiftlet *Collocalia brevirostris*	m 5
R	White-rumped Needletail *Zoonavena sylvatica*	r? 3
	White-throated Needletail *Hirundapus caudacutus*	m 3
I	White-vented Needletail *Hirundapus cochinchinensis*	r? 4
	Pacific Swift *Apus pacificus*	w? 5
	Alpine Swift *Apus melba*	r? 2
	Little Swift *Apus affinis*	r? m? 5
	Asian Palm Swift *Cypsiurus balasiensis*	m 5
	Crested Tree Swift *Hemiprocne coronata*	r 1
E	Red-headed Trogon *Harpactes erythrocephalus*	r 5
	White-breasted Kingfisher *Halcyon smyrnensis*	r 1
	Black-capped Kingfisher *Halcyon pileata*	v
	Stork-billed Kingfisher *Pelargopsis capensis*	r 3
	Common Kingfisher *Alcedo atthis*	r 2
E	Deep-blue Kingfisher *Alcedo meninting*	r 4
	Pied Kingfisher *Ceryle rudis*	r 1
	Crested Kingfisher *Ceryle lugubris*	r? 5
	Blue-bearded Bee-eater *Nyctyornis athertoni*	r 4
	Green Bee-eater *Merops orientalis*	r s 1
	Blue-tailed Bee-eater *Merops philippinus*	s 2
	Chestnut-headed Bee-eater *Merops leschenaulti*	r 5 s 1
	Indian Roller *Coracias benghalensis*	r 1
R	Dollarbird *Eurystomus orientalis*	s 1
	Hoopoe *Upupa epops*	r? 2
	Indian Grey Hornbill *Tockus birostris*	r 5
V	Oriental Pied Hornbill *Anthracoceros coronatus*	r 2
E	Great Pied Hornbill *Buceros bicornis*	r 4
	Lineated Barbet *Megalaima lineata*	r 1
	Blue-throated Barbet *Megalaima asiatica*	r 3
	Coppersmith Barbet *Megalaima haemacephala*	r 4
	Eurasian Wryneck *Jynx torquilla*	w m 3
	Speckled Piculet *Picumnus innominatus*	r 5
R	White-browed Piculet *Sasia ochracea*	r 4
	Rufous Woodpecker *Celeus brachyurus*	r 4
	Lesser Yellow-naped Woodpecker *Picus chlorolophus*	r 2
	Greater Yellow-naped Woodpecker *Picus flavinucha*	r 3
	Grey-headed Woodpecker *Picus canus*	r 2

	Streak-throated Green Woodpecker *Picus myrmecophoneus*	r 2
	Himalayan Golden-backed Woodpecker *Dinopium shorii*	r 1
	Lesser Golden-backed Woodpecker *Dinopium benghalense*	r 3
	Greater Golden-backed Woodpecker *Chrysocolaptes lucidus*	r 2
E	Great Slaty Woodpecker *Mulleripicus pulverulentus*	r 5
	Yellow-crowned Pied Woodpecker *Dendrocopos mahrattensis*	r 5
	Fulvous-breasted Pied Woodpecker *Dendrocopos macei*	r 2
	Grey-capped Pygmy Woodpecker *Dendrocopos canicapillus*	r 1
	Brown-capped Pygmy Woodpecker *Dendrocopos moluccensis*	r 5
E	Long-tailed Broadbill *Psarisomus dalhousiae*	r 5
R	Hooded Pitta *Pitta sordida*	s 1
R	Indian Pitta *Pitta brachyura*	s 1
	Bengal Bush Lark *Mirafra assamica*	r 1
	Ashy-crowned Finchlark *Eremopterix grisea*	r 3
	Greater Short-toed Lark *Calandrella brachydactyla*	w? m? 5
	Sandlark *Calandrella raytal*	r 1
	Oriental Skylark *Alauda gulgula*	r 3
	Brown-throated Sand Martin *Riparia paludicola*	r 1
	Collared Sand Martin *Riparia riparia*	v
	Barn Swallow *Hirundo rustica*	r 2
	Red-rumped Swallow *Hirundo daurica*	r 2
	Nepal House-Martin *Delichon nipalensis*	w 5
	Asian House-Martin *Delichon dasypus*	v
	Common House-Martin *Delichon urbica*	v
	Richard's Pipit *Anthus novaeseelandiae*	r 1
	Tawny Pipit *Anthus campestris*	v
	Olive-backed Pipit *Anthus hodgsoni*	w 1
	Tree Pipit *Anthus trivialis*	w m 4
	Red-throated Pipit *Anthus cervinus*	v
	Rosy Pipit *Anthus roseatus*	w m 2
	Buff-bellied/Water Pipit *Anthus rubescens/spinoletta*	v
	Forest Wagtail *Dendronanthus indicus*	v
	Yellow Wagtail *Motacilla flava*	w m 2
	Citrine Wagtail *Motacilla citreola*	w m 2
	Grey Wagtail *Motacilla cinerea*	w 2
	White Wagtail *Motacilla alba*	w m 1
	White-browed Wagtail *Motacilla maderaspatensis*	r 1
	Common Woodshrike *Tephrodornis pondicerianus*	r 2
	Large Woodshrike *Tephrodornis gularis*	r 4
	Bar-winged Flycatcher-shrike *Hemipus picatus*	r 2
	Black-headed Cuckoo-shrike *Coracina melanoptera*	m 5
	Black-winged Cuckoo-shrike *Coracina melaschistos*	r 2
	Large Cuckoo-shrike *Coracina novaehollandiae*	r 1
	Scarlet Minivet *Pericrocotus flammeus*	r 1
	Long-tailed Minivet *Pericrocotus ethologus*	w 3
	Small Minivet *Pericrocotus cinnamomeus*	r 3
R	Rosy Minivet *Pericrocotus roseus*	r? 4
	Black-crested Bulbul *Pycnonotus melanicterus*	r 3
	Red-whiskered Bulbul *Pycnonotus jocosus*	r 1
	White-cheeked Bulbul *Pycnonotus leucogenys*	r 3
	Red-vented Bulbul *Pycnonotus cafer*	r 1
V	White-throated Bulbul *Criniger flaveolus*	r? 5

	Black Bulbul *Hypsipetes madagascariensis*	r 3
	Common Iora *Aegithina tiphia*	r 1
	Golden-throated Leafbird *Chloropsis aurifrons*	r 2
	Orange-bellied Leafbird *Chloropsis hardwickii*	r 4
R	White-browed Shortwing *Brachypteryx montana*	w 5
	Siberian Rubythroat *Luscinia calliope*	w m 3
	Bluethroat *Luscicnia svecica*	w m 2
	White-tailed Rubythroat *Luscinia pectoralis*	w 3
	Indian Blue Robin *Luscinia brunnea*	m 5
	Asian Magpie-Robin *Copsychus saularis*	r 1
	White-rumped Shama *Copsychus malabaricus*	r 1
	Black Redstart *Phoenicurus ochruros*	w 3
	Plumbeous Redstart *Rhyacornis fuliginosus*	w 4
R	White-tailed Robin *Cinclidium leucurum*	w? 5
	Common Stonechat *Saxicola torquata*	r 1
R+	White-tailed Stonechat *Saxicola leucura*	r 2
	Hodgson's Bushchat *Saxicola insignis*	v
	Pied Bushchat *Saxicola caprata*	r 1
	Grey Bushchat *Saxicola ferrea*	w 4
	Northern Wheatear *Oenanthe oenanthe*	v
	Desert Wheatear *Oenanthe deserti*	v
	White-capped Redstart *Chaimarrornis leucocephalus*	w 5
	Indian Robin *Saxicoloides fulicata*	r 5
	Blue-capped Rock-Thrush *Monticola cinclorhyncha*	m 5
	Blue Rock-Thrush *Monticola solitarius*	w 4
	Blue Whistling Thrush *Myiophoneus caeruleus*	r 3
	Scaly Thrush *Zoothera dauma*	w 5
R*	Long-billed Thrush *Zoothera monticola*	w 5
	Orange-headed Thrush *Zoothera citrina*	s 1
+	Tickell's Thrush *Turdus unicolor*	w 5
+	Grey-winged Blackbird *Turdus boulboul*	w 4
	Dark-throated Thrush *Turdus ruficollis*	w m 1
+	Black-backed Forktail *Enicurus immaculatus*	r 3
	Chestnut-headed Tesia *Tesia castaneocoronata*	w 5
	Grey-bellied Tesia *Tesia cyaniventer*	w 5
R*	Pale-footed Bush Warbler *Cettia pallidipes*	r 2
*	Chestnut-crowned Bush Warbler *Cettia major*	w 5
+	Aberrant Bush Warbler *Cettia flavolivacea*	w 5
+	Grey-sided Bush Warbler *Cettia brunnifrons*	w 5
	Spotted Bush Warbler *Bradypterus thoracicus*	w 5
R	Bright-capped Cisticola *Cisticola exilis*	r 2
	Fantail Cisticola *Cisticola juncidis*	r 2
	Graceful Prinia *Prinia gracilis*	r 5
	Plain Prinia *Prinia inornata*	r 3
	Ashy Prinia *Prinia socialis*	r 3
	Grey-breasted Prinia *Prinia hodgsoni*	r 2
	Yellow-bellied Prinia *Prinia flaviventris*	r 2
	Striated Prinia *Prinia criniger*	r? w? 4
*	Grey-capped Prinia *Prinia cinereocapilla*	r 2
V	Large Grass Warbler *Graminicola bengalensis*	r 3
	Common Tailorbird *Orthotomus sutorius*	r 1
	Lanceolated Warbler *Locustella lanceolata*	v

	Common Grasshopper Warbler *Locustella naevia*	v
I	Bristled Grass Warbler *Chaetornis striatus*	s? 5
V	Striated Marsh Warbler *Megalurus palustris*	r 5
	Paddyfield Warbler *Acrocephalus agricola*	w 5
	Blyth's Reed Warbler *Acrocephalus dumetorum*	w m 2
	Thick-billed Warbler *Acrocephalus aedon*	w 5
	Booted Warbler *Hippolais caligata*	v
	Orphean Warbler *Sylvia hortensis*	v
	Lesser Whitethroat *Sylvia curruca*	w m 4
	Golden-spectacled Warbler *Seicercus burkii*	w 3
	Chestnut-crowned Warbler *Seicercus castaniceps*	w 4
+	Grey-hooded Warbler *Seicercus xanthoschistos*	w 4
	Yellow-bellied Warbler *Abroscopus superciliaris*	w 4
	Blyth's Crowned Warbler *Phylloscopus reguloides*	w 2
	Western Crowned Warbler *Phylloscopus occipitalis*	m 5
	Green Warbler *Phylloscopus nitidus*	m 5
	Greenish Warbler *Phylloscopus trochiloides*	w m 1
	Large-billed Leaf Warbler *Phylloscopus magnirostris*	m 5
	Pallas's Leaf Warbler *Phylloscopus proregulus*	w 4
	Yellow-browed Warbler *Phylloscopus inornatus*	w 2
	Dusky Warbler *Phylloscopus fuscatus*	w 3
*	Smoky Warbler *Phylloscopus fuligiventer*	w 4
	Sulphur-bellied Warbler *Phylloscopus griseolus*	v
	Tickell's Warbler *Phylloscopus affinis*	w 3
	Chiffchaff *Phylloscopus collybita*	w 3
	Pale-chinned Flycatcher *Cyornis poliogenys*	r 1
	Blue-throated Blue Flycatcher *Cyornis rubeculoides*	w 5
	Tickell's Blue Flycatcher *Cyornis tickelliae*	? 5
	Verditer Flycatcher *Muscicapa thalassina*	w 3
	Asian Sooty Flycatcher *Muscicapa sibirica*	m 5
	Rufous-tailed Flycatcher *Muscicapa ruficauda*	m 5
	Asian Brown Flycatcher *Muscicapa latirostris*	w m 4
	Slaty-blue Flycatcher *Ficedula tricolor*	w 5
	Ultramarine Flycatcher *Ficedula superciliaris*	m 5
R	Little Pied Flycatcher *Ficedula westermanni*	w 4
	Slaty-backed Flycatcher *Ficedula hodgsonii*	w 5
	Snowy-browed Flycatcher *Ficedula hyperythra*	w 5
	Orange-gorgetted Flycatcher *Ficedula strophiata*	w 5
	Kashmir Flycatcher *Ficedula subrubra*	m 5
	Red-breasted Flycatcher *Ficedula parva*	w 1
	Grey-headed Flycatcher *Culicicapa ceylonensis*	r 2
+	Yellow-bellied Fantail *Rhipidura hypoxantha*	w 5
	White-throated Fantail *Rhipidura albicollis*	r 1
	White-browed Fantail *Rhipidura aureola*	r 3
	Asian Paradise Flycatcher *Terpsiphone paradisi*	s 1
	Black-naped Monarch *Hypothymis azurea*	r 3
	Puff-throated Babbler *Pellorneum ruficeps*	r 1
	Rusty-cheeked Scimitar-Babbler *Pomatorhinus erythrogenys*	r 4
	White-browed Scimitar-Babbler *Pomatorhinus schisticeps*	r 4
	Lesser Scaly-breasted Wren-Babbler *Pnoepyga pusilla*	w 4
+	Black-chinned Babbler *Stachyris pyrrhops*	r 2
	Grey-throated Babbler *Stachyris nigriceps*	r w 3

	Rufous-bellied Babbler *Dumetia hyperythra*	r 5
	Striped Tit-Babbler *Macronous gularis*	r 1
	Red-capped Babbler *Timalia pileata*	r 1
	Yellow-eyed Babbler *Chrysomma sinense*	r 2
	Striated Babbler *Turdoides earlei*	r 1
V+	Slender-billed Babbler *Turdoides longirostris*	r 2
	Jungle Babbler *Turdoides striatus*	r 2
V	Lesser Necklaced Laughing-thrush *Garrulax monileger*	r 3
V	Greater Necklaced Laughing-thrush *Garrulax pectoralis*	r 4
V+	Rufous-necked Laughing-thrush *Garrrulax ruficollis*	r 2
+	Nepal Fulvetta *Alcippe nipalensis*	w? 4
	White-bellied Yuhina *Yuhina zantholeuca*	r 3
	Great Tit *Parus major*	r 1
	Velvet-fronted Nuthatch *Sitta frontalis*	r 1
	Chestnut-bellied Nuthatch *Sitta castanea*	r 1
	Wallcreeper *Tichodroma muraria*	w 4
E	Ruby-cheeked Sunbird *Anthreptes singalensis*	r 5
	Purple Sunbird *Nectarinia asiatica*	r 1
	Black-throated Sunbird *Aethopyga saturata*	m 5
	Crimson Sunbird *Aethopyga siparaja*	r 2
E	Little Spiderhunter *Arachnothera longirostra*	r 5
	Streaked Spiderhunter *Arachnothera magna*	r 4
	Thick-billed Flowerpecker *Dicaeum agile*	r 4
	Pale-billed Flowerpecker *Dicaeum erythrorhynchos*	r 3
	Plain Flowerpecker *Dicaeum concolor*	? 5
	Buff-bellied Flowerpecker *Dicaeum ignipectus*	w 5
	Oriental White-eye *Zosterops palpebrosa*	r 1
	Maroon Oriole *Oriolus traillii*	v
	Black-hooded Oriole *Oriolus xanthornus*	r 1
	Slender-billed Oriole *Oriolus tenuirostris*	v
	Eurasian Golden Oriole *Oriolus oriolus*	s 1
	Brown Shrike *Lanius cristatus*	w m 5
	Bay-backed Shrike *Lanius vittatus*	v
	Long-tailed Shrike *Lanius schach*	r 3
	Grey-backed Shrike *Lanius tephronotus*	w 4
	Black Drongo *Dicrurus macrocercus*	r 2
	Ashy Drongo *Dicrurus leucophaeus*	r 2
	White-bellied Drongo *Dicrurus caerulescens*	r 2
V	Crow-billed Drongo *Dicrurus annectans*	s 4
	Bronzed Drongo *Dicrurus aeneus*	r 2
	Lesser Racket-tailed Drongo *Dicrurus remifer*	r 4
	Spangled Drongo *Dicrurus hottentottus*	r 2
	Greater Racket-tailed Drongo *Dicrurus paradiseus*	r 3
	Ashy Woodswallow *Artamus fuscus*	r 2
	Red-billed Blue Magpie *Urocissa erythrorhyncha*	r 3
	Green Magpie *Cissa chinensis*	r 4
	Rufous Treepie *Dendrocitta vagabunda*	r 1
	House Crow *Corvus splendens*	r 5
	Jungle Crow *Corvus macrorhynchos*	r 1
	Spot-winged Stare *Saroglossa spiloptera*	m 2
	Chestnut-tailed Starling *Sturnus malabaricus*	r? s? 2
	Brahminy Starling *Sturnus pagodarum*	r? 5

	Common Starling *Sturnus vulgaris*	m 5
	Asian Pied Starling *Sturnus contra*	r 3
	Common Mynah *Acridotheres tristis*	r 3
	Bank Mynah *Acridotheres ginginianus*	r 2
	Jungle Mynah *Acridotheres fuscus*	r 1
	Hill Mynah *Gracula religiosa*	r 4
	House Sparrow *Passer domesticus*	w 3
	Yellow-throated Sparrow *Petronia xanthocollis*	r 3
R+	Black-breasted Weaver *Ploceus benghalensis*	r 2
	Baya Weaver *Ploceus philippinus*	r 1
	Red Avadavat *Amandava amandava*	r 3
	Indian Silverbill *Euodice malabarica*	m 5
	Striated Munia *Lonchura striata*	r 3
	Scaly-breasted Munia *Lonchura punctulata*	r 1
	Chestnut Munia *Lonchura malacca*	r 2
	Common Rosefinch *Carpodacus erythrinus*	w 2
	Black-faced Bunting *Emberiza spodocephala*	w m 4
	Chestnut-eared Bunting *Emberiza fucata*	w 5
	Rustic Bunting *Emberiza rustica*	v
	Yellow-breasted Bunting *Emberiza aureola*	w 1
	Red-headed Bunting *Emberiza bruniceps*	v
	Crested Bunting *Melophus lathami*	w 3

LANGTANG NATIONAL PARK

	Bar-headed Goose *Anser indicus*	m 5
	Ruddy Shelduck *Tadorna ferruginea*	s b 4
	Common Teal *Anas crecca*	m 5
	Tufted Duck *Aythya fuligula*	m 5
	Crested Honey Buzzard *Pernis ptilorhyncus*	r? 5
	Black Kite *Milvus migrans*	r? 4
	Lammergeier *Gypaetus barbatus*	r 1
	Himalayan Griffon Vulture *Gyps himalayensis*	r 2
	Red-headed Vulture *Sarcogyps calvus*	r? 3
	Crested Serpent Eagle *Spilornis cheela*	s 2
	Hen Harrier *Circus cyaneus*	w m 2
	Pallid Harrier *Circus macrourus*	w m 5
	Northern Goshawk *Accipiter gentilis*	r 3
	Besra *Accipiter virgatus*	r 4
	Northern Sparrowhawk *Accipiter nisus*	s w 3
	Common Buzzard *Buteo buteo*	w m 3
	Long-legged Buzzard *Buteo rufinus*	w m 4
	Upland Buzzard *Buteo hemilasius*	?
	Black Eagle *Ictinaetus malayensis*	r 2
	Steppe Eagle *Aquila rapax nipalensis*	w m 2
	Imperial Eagle *Aquila heliaca*	w m 5
	Golden Eagle *Aquila chrysaetos*	r 2
	Bonelli's Eagle *Hieraaetus fasciatus*	r 2
	Mountain Hawk-Eagle *Spizaetus nipalensis*	r 3
	Common Kestrel *Falco tinnunculus*	r w m 1
	Peregrine *Falco peregrinus*	r 4
+	Snow Partridge *Lerwa lerwa*	r 2
	Tibetan Snowcock *Tetraogallus tibetanus*	r 2
	Himalayan Snowcock *Tetraogallus himalayensis*	r? 5
	Chukar Partridge *Alectoris chukar*	r 4
	Black Francolin *Francolinus francolinus*	s 4
	Common Quail *Coturnix coturnix*	s 5
	Common Hill Partridge *Arborophila torqueola*	r 3
+	Blood Pheasant *Ithaginis cruentus*	r 2
V+	Satyr Tragopan *Tragopan satyra*	r 4
+	Himalayan Monal *Lophophorus impejanus*	r b 1
	Kalij Pheasant *Lophura leucomelana*	r 2
	Ibisbill *Ibidorhyncha struthersii*	s b 2
	Eurasian Woodcock *Scolopax rusticola*	s b 3
	Eurasian Curlew *Numenius arquata*	v
	Rock Pigeon *Columba livia*	r 1
	Snow Pigeon *Columba leuconota*	r 1
+	Speckled Woodpigeon *Columba hodgsonii*	r 2
+	Ashy Woodpigeon *Columba pulchricollis*	r 4
	Oriental Turtle Dove *Streptopelia orientalis*	s 1
	Spotted Dove *Streptopelia chinensis*	s 4
	Wedge-tailed Green Pigeon *Treron sphenura*	r 2
	Large Hawk-Cuckoo *Hierococcyx sparverioides*	s 2
	Indian Cuckoo *Cuculus micropterus*	s 3
	Common Cuckoo *Cuculus canorus*	s 1

	Oriental Cuckoo *Cuculus saturatus*	s 1
	Lesser Cuckoo *Cuculus poliocephalus*	s 2
	Common Koel *Eudynamys scolopacea*	s 5
R	Mountain Scops Owl *Otus spilocephalus*	r 5
	Collared Owlet *Glaucidium brodiei*	r 3
	Spotted Little Owl *Athene brama*	v
	Tawny Owl *Strix aluco*	r 4
	Short-eared Owl *Asio flammeus*	m 5
	Jungle Nightjar *Caprimulgus indicus*	r? 2
	Himalayan Swiftlet *Collocalia brevirostris*	r? 2
	White-throated Needletail *Hirundapus caudacutus*	? 3
	Pacific Swift *Apus pacificus*	s b 2
	Alpine Swift *Apus melba*	r? 3
	Little Swift *Apus affinis*	s 3
	Hoopoe *Upupa epops*	s m 4
	Great Barbet *Megalaima virens*	r 1
	Golden-throated Barbet *Megalaima franklinii*	r 3
	Blue-throated Barbet *Megalaima asiatica*	r 3
R*	Orange-rumped Honeyguide *Indicator xanthonotus*	r 3
	Speckled Piculet *Picumnus innominatus*	r? 4
	Grey-headed Woodpecker *Picus canus*	r? 4
	Scaly-bellied Green Woodpecker *Picus squamatus*	r 1
V	Bay Woodpecker *Blythipicus pyrrhotis*	r 5
+	Darjeeling Pied Woodpecker *Dendrocopos darjellensis*	r 2
+	Crimson-breasted Pied Woodpecker *Dendrocopos cathpharius*	r 3
	Rufous-bellied Pied Woodpecker *Dendrocopos hyperythrus*	r 3
+	Brown-fronted Pied Woodpecker *Dendrocopos auriceps*	r 1
	Fulvous-breasted Pied Woodpecker *Dendrocopos macei*	r 5
	Crag Martin *Ptyonoprogne rupestris*	s? 5
	Barn Swallow *Hirundo rustica*	s 4
	Red-rumped Swallow *Hirundo daurica*	s 4
+	Nepal House-Martin *Delichon nipalensis*	r? b 2
	Asian House-Martin *Delichon dasypus*	? b 3
	Blyth's Pipit *Anthus godlewskii*	m 5
	Olive-backed Pipit *Anthus hodgsoni*	r 1
	Red-throated Pipit *Anthus cervinus*	m 5
	Rosy Pipit *Anthus roseatus*	s 1
	Water Pipit *Anthus spinoletta*	m 5
	Upland Pipit *Anthus sylvanus*	r 2
	Yellow Wagtail *Motacilla flava*	m 5
	Grey Wagtail *Motacilla cinerea*	s 2
	White Wagtail *Motacilla alba*	s m 2
	Black-winged Cuckoo-shrike *Coracina melaschistos*	s 4
	Large Cuckoo-shrike *Coracina novaehollandiae*	s? 4
	Scarlet Minivet *Pericrocotus flammeus*	r? 3
R	Short-billed Minivet *Pericrocotus brevirostris*	r? 5
	Long-tailed Minivet *Pericrocotus ethologus*	r 1
+	Striated Bulbul *Pycnonotus striatus*	r? 4
	White-cheeked Bulbul *Pycnonotus leucogenys*	r 1
	Mountain Bulbul *Hypsipetes mcclellandii*	r 2
	Black Bulbul *Hypsipetes madagascariensis*	r 1
	Orange-bellied Leafbird *Chloropsis hardwickii*	r 3

	Bohemian Waxwing *Bombycilla garrulus*	v
	White-breasted Dipper *Cinclus cinclus*	r 4
	Brown Dipper *Cinclus pallasii*	r b 1
	Northern Wren *Troglodytes troglodytes*	r 1
	Maroon-backed Accentor *Prunella immaculata*	w 4
+	Rufous-breasted Accentor *Prunella strophiata*	s 2
	Brown Accentor *Prunella fulvescens*	v
+	Robin Accentor *Prunella rubeculoides*	r 3
	Altai Accentor *Prunella himalayana*	w 3
	Alpine Accentor *Prunella collaris*	r 2
R*	Gould's Shortwing *Brachypteryx stellata*	l s 2
R	White-browed Shortwing *Brachypteryx montana*	l s 3
	White-tailed Rubythroat *Luscinia pectoralis*	s b 3
+	Indian Blue Robin *Luscinia brunnea*	s 2
	Orange-flanked Bush-Robin *Tarsiger cyanurus*	r 1
+	Golden Bush-Robin *Tarsiger chrysaeus*	r b 2
+	White-browed Bush-Robin *Tarsiger indicus*	r 3
*	Rufous-breasted Bush-Robin *Tarsiger hyperythrus*	r b 2
	Blue-capped Redstart *Phoenicurus caeruleocephalus*	s 3
	Black Redstart *Phoenicurus ochruros*	s 3
	Hodgson's Redstart *Phoenicurus hodgsoni*	w 4
+	Blue-fronted Redstart *Phoenicurus frontalis*	r b 1
+	White-throated Redstart *Phoenicurus schisticeps*	r 4
	Güldenstadt's Redstart *Phoenicurus erythrogaster*	r? w 4
	Plumbeous Redstart *Rhyacornis fuliginosus*	r 1
+	White-bellied Redstart *Hodgsonius phoenicuroides*	s 3
R	White-tailed Robin *Cinclidium leucurum*	r? 5
+	Grandala *Grandala coelicolor*	r 1
	Common Stonechat *Saxicola torquata*	r 2
	Pied Bushchat *Saxicola caprata*	s 3
	Grey Bushchat *Saxicola ferrea*	r? 2
	White-capped Redstart *Chaimarrornis leucocephalus*	r 1
+	Blue-capped Rock-Thrush *Monticola cinclorhyncha*	s b 2
	Chestnut-bellied Rock-Thrush *Monticola rufiventris*	r 2
	Blue Whistling Thrush *Myiophoneus caeruleus*	r 1
+	Plain-backed Mountain Thrush *Zoothera mollissima*	r 2
+	Long-tailed Mountain Thrush *Zoothera dixoni*	r 3
	Scaly Thrush *Zoothera dauma*	s b 3
R*	Long-billed Thrush *Zoothera monticola*	r 3
*	Pied Ground Thrush *Zoothera wardii*	s b 2
+	Tickell's Thrush *Turdus unicolor*	s 3
+	White-collared Blackbird *Turdus albocinctus*	r 2
+	Grey-winged Blackbird *Turdus boulboul*	r? 3
	Eurasian Blackbird *Turdus merula*	m 5
	Eye-browed Thrush *Turdus obscurus*	v
	Dark-throated Thrush *Turdus ruficollis*	w 1
	Little Forktail *Enicurus scouleri*	s 3
	Spotted Forktail *Enicurus maculatus*	r 3
+	Chestnut-headed Tesia *Tesia castaneocoronata*	r 2
	Grey-bellied Tesia *Tesia cyaniventer*	s 3
+	Aberrant Bush Warbler *Cettia flavolivacea*	s b 3
+	Grey-sided Bush Warbler *Cettia brunnifrons*	r? 1

	Striated Prinia *Prinia criniger*	r 1
	Golden-spectacled Warbler *Seicercus burkii*	r 1
	Chestnut-crowned Warbler *Seicercus castaniceps*	r 3
+	Grey-hooded Warbler *Seicercus xanthoschistos*	r 1
+	Black-faced Warbler *Abroscopus schisticeps*	r 2
	Blyth's Crowned Warbler *Phylloscopus reguloides*	s 1
	Greenish Warbler *Phylloscopus trochiloides*	s 1
+	Large-billed Leaf Warbler *Phylloscopus magnirostris*	s 3
+	Orange-barred Leaf Warbler *Phylloscopus pulcher*	r 1
	Grey-faced Leaf Warbler *Phylloscopus maculipennis*	r 2
	Pallas's Leaf Warbler *Phylloscopus proregulus*	r 1
	Yellow-browed Warbler *Phylloscopus inornatus*	w m 3
*	Smoky Warbler *Phylloscopus fuligiventer*	s 4
	Tickell's Warbler *Phylloscopus affinis*	s 1
	Goldcrest *Regulus regulus*	r 3
V	Large Niltava *Niltava grandis*	r 5
	Small Niltava *Niltava macgrigoriae*	s 4
+	Rufous-bellied Niltava *Niltava sundara*	r? b 2
	Verditer Flycatcher *Muscicapa thalassina*	s b 1
R	Ferruginous Flycatcher *Muscicapa ferruginea*	s 4
	Asian Sooty Flycatcher *Muscicapa sibirica*	s 1
+	Rufous-tailed Flycatcher *Muscicapa ruficauda*	s 4
	Asian Brown Flycatcher *Muscicapa latirostris*	s 5
	Slaty-blue Flycatcher *Ficedula tricolor*	r? 2
+	Ultramarine Flycatcher *Ficedula superciliaris*	s 2
R	Little Pied Flycatcher *Ficedula westermanni*	s 4
	Snowy-browed Flycatcher *Ficedula hyperythra*	s 4
	Orange-gorgetted Flycatcher *Ficedula strophiata*	r 1
	Red-breasted Flycatcher *Ficedula parva*	w 3
	Grey-headed Flycatcher *Culicicapa ceylonensis*	s 1
+	Yellow-bellied Fantail *Rhipidura hypoxantha*	r? 1
	White-throated Fantail *Rhipidura albicollis*	r? 3
	Rusty-cheeked Scimitar-Babbler *Pomatorhinus erythrogenys*	r 3
	Streak-breasted Scimitar-Babbler *Pomatorhinus ruficollis*	r 3
+	Greater Scaly-breasted Wren-Babbler *Pnoepyga albiventer*	r 2
	Lesser Scaly-breasted Wren-Babbler *Pnoepyga pusilla*	r? 3
+	Black-chinned Babbler *Stachyris pyrrhops*	r 4
V*	Great Parrotbill *Conostoma aemodium* (recorded between 1947 and 1961, locally very common in 1960/61 (Proud 1961), but no subsequent records, confirmation of continued presence needed)	?
V*	Fulvous Parrotbill *Paradoxornis fulvifrons*	r 5
R	Black-throated Parrotbill *Paradoxornis nipalensis*	r 5
+	White-throated Laughing-thrush *Garrulax albogularis*	r 1
+	Striated Laughing-thrush *Garrulax striatus*	r 1
+	Variegated Laughing-thrush *Garrulax variegatus*	r 1
+	Spotted Laughing-thrush *Garrulax ocellatus*	r 2
	Streaked Laughing-thrush *Garrulax lineatus*	r 1
V+	Scaly Laughing-thrush *Garrulax subunicolor*	r 4
+	Black-faced Laughing-thrush *Garrulax affinis*	r 1
	Chestnut-crowned Laughing-thrush *Garrulax erythrocephalus*	r 1
R*	Fire-tailed Myzornis *Myzornis pyrrhoura*	r b 2

	White-browed Shrike-Babbler *Pteruthius flaviscapis*	r 2
+	Green Shrike-Babbler *Pteruthius xanthochloris*	r 3
*	Hoary Barwing *Actinodura nipalensis*	r 2
	Chestnut-tailed Minla *Minla strigula*	r 1
	Rufous-winged Fulvetta *Alcippe castaneceps*	r 3
+	White-browed Fulvetta *Alcippe vinipectus*	r 1
+	Black-capped Sibia *Heterophasia capistrata*	r 1
+	Whiskered Yuhina *Yuhina flavicollis*	r 1
+	Stripe-throated Yuhina *Yuhina gularis*	r 1
+	Rufous-vented Yuhina *Yuhina occipitalis*	r 1
+	Black-browed Tit *Aegithalos iouschistos*	r b 2
	White-throated Tit *Aegithalos niveogularis*	v
	Black-throated Tit *Aegithalos concinnus*	r b 1
	Yellow-browed Tit *Sylviparus modestus*	r 2
+	Grey-crested Tit *Parus dichrous*	r 1
+	Rufous-vented Black Tit *Parus rubidiventris*	r 1
	Coal Tit *Parus ater*	r 1
	Great Tit *Parus major*	r? 4
	Green-backed Tit *Parus monticolus*	r 2
	Black-lored Tit *Parus xanthogenys*	r 2
+	White-tailed Nuthatch *Sitta himalayensis*	r 1
	Wallcreeper *Tichodroma muraria*	w 2
*	Rusty-flanked Treecreeper *Certhia nipalensis*	r b 2
	Common Treecreeper *Certhia familiaris*	r 2
+	Fire-capped Tit *Cephalopyrus flammiceps*	? 5
	Mrs Gould's Sunbird *Aethopyga gouldiae*	r b 3
	Green-tailed Sunbird *Aethopyga nipalensis*	r 1
+	Fire-tailed Sunbird *Aethopyga ignicauda*	r 1
	Thick-billed Flowerpecker *Dicaeum agile*	v
R*	Yellow-bellied Flowerpecker *Dicaeum melanoxanthum*	?4
	Buff-bellied Flowerpecker *Dicaeum ignipectus*	r 1
	Oriental White-eye *Zosterops palpebrosa*	s 3
	Maroon Oriole *Oriolus traillii*	s 4
	Golden Oriole *Oriolus oriolus*	s 4
	Long-tailed Shrike *Lanius schach*	r 1
+	Grey-backed Shrike *Lanius tephronotus*	r 2
	Black Drongo *Dicrurus macrocercus*	s 4
	Ashy Drongo *Dicrurus leucophaeus*	s 2
	Bronzed Drongo *Dicrurus aeneus*	s 4
	Eurasian Jay *Garrulus glandarius*	r 3
+	Lanceolated Jay *Garrulus lanceolatus*	r 4
+	Yellow-billed Blue Magpie *Urocissa flavirostris*	r 1
	Grey Treepie *Dendrocitta formosae*	s 4
	Eurasian Nutcracker *Nucifraga caryocatactes*	r 1
	Alpine Chough *Pyrrhocorax graculus*	r 1
	Red-billed Chough *Pyrrhocorax pyrrhocorax*	r 1
	Jungle Crow *Corvus macrorhynchos*	r 1
	Common Raven *Corvus corax*	r 3
	Common Mynah *Acridotheres tristis*	r? 3
	House Sparrow *Passer domesticus*	r 5
	Eurasian Tree Sparrow *Passer montanus*	r 1
	Tibetan Snowfinch *Montifringilla adamsi*	w?

	Common Chaffinch *Fringilla coelebs*	w 5
	Tibetan Serin *Serinus thibetanus*	w 4
	Yellow-breasted Greenfinch *Carduelis spinoides*	s 1
	Eurasian Goldfinch *Carduelis carduelis*	r 4
	Twite *Carduelis flavirostris*	r? 4
	Common Crossbill *Loxia curvirostra*	r? 4
	Plain Mountain-Finch *Leucosticte nemoricola*	r 1
	Crimson Rosefinch *Carpodacus rubescens*	v
+	Dark-breasted Rosefinch *Carpodacus nipalensis*	r 1
	Common Rosefinch *Carpodacus erythrinus*	s 1
	Beautiful Rosefinch *Carpodacus pulcherrimus*	r 1
+	Pink-browed Rosefinch *Carpodacus rhodochrous*	r 1
*	Vinaceous Rosefinch *Carpodacus vinaceus*	? 5
+	Dark-rumped Rosefinch *Carpodacus edwardsii*	? 5
+	Spot-winged Rosefinch *Carpodacus rhodopeplus*	r 2
+	White-browed Rosefinch *Carpodacus thura*	r 2
	Red-breasted Rosefinch *Carpodacus puniceus*	r 3
*	Crimson-browed Finch *Propyrrhula subhimachala*	r 3
V*	Scarlet Finch *Haematospiza sipahi*	r 3
+	Gold-naped Finch *Pyrrhoplectes epauletta*	r 5
+	Red-headed Bullfinch *Pyrrhula erythrocephala*	r 1
	Brown Bullfinch *Pyrrhula nipalensis*	r 4
+	Collared Grosbeak *Mycerobas affinis*	r 2
*	Spot-winged Grosbeak *Mycerobas melanozanthos*	r? 5
	White-winged Grosbeak *Mycerobas carnipes*	r 2
	Little Bunting *Emberiza pusilla*	w 4
	Crested Bunting *Melophus lathami*	s 3

SAGARMATHA NATIONAL PARK

	Bar-headed Goose *Anser indicus*	m 3
	Ruddy Shelduck *Tadorna ferruginea*	r 3
	Eurasian Wigeon *Anas penelope*	m 4
	Gadwall *Anas strepera*	m 5
	Northern Pintail *Anas acuta*	m 5
	Northern Shoveler *Anas clypeata*	m 5
	Common Pochard *Aythya ferina*	m 5
	Ferruginous Duck *Aythya nyroca*	m 5
	Tufted Duck *Aythya fuligula*	m 4
	Common Goldeneye *Bucephala clangula*	m 5
	Black Kite *Milvus migrans*	s 4
	Lesser Fishing Eagle *Ichthyophaga nana*	v
	Lammergeier *Gypaetus barbatus*	r 4
	Himalayan Griffon Vulture *Gyps himalayensis*	r 2
	Hen Harrier *Circus cyaneus*	m 5
	Northern Goshawk *Accipiter gentilis*	r 3
	Besra *Accipiter virgatus*	v
	Northern Sparrowhawk *Accipiter nisus*	r 3
	Common Buzzard *Buteo buteo*	w m 3
	Long-legged Buzzard *Buteo rufinus*	m 5
	Steppe Eagle *Aquila rapax nipalensis*	m 5
	Imperial Eagle *Aquila heliaca*	m 5
	Golden Eagle *Aquila chrysaetos*	r 4
	Osprey *Pandion haliaetus*	m 5
	Common Kestrel *Falco tinnunculus*	r? 3
	Merlin *Falco columbarius*	v
	Peregrine *Falco peregrinus*	r 4
	Barbary Falcon *Falco pelegrinoides*	v
+	Snow Partridge *Lerwa lerwa*	r 1
	Tibetan Snowcock *Tetraogallus tibetanus*	r 1
+	Blood Pheasant *Ithaginis cruentus*	r 2
V+	Satyr Tragopan *Tragopan satyra*	r 5
+	Himalayan Monal *Lophophorus impejanus*	r 1
	Kalij Pheasant *Lophura leucomelana*	s 4
	Common Coot *Fulica atra*	m 5
	Demoiselle Crane *Anthropoides virgo*	m 5
	Ibisbill *Ibidorhyncha struthersii*	s b 4
	Common Snipe *Gallinago gallinago*	m 5
	Eurasian Woodcock *Scolopax rusticola*	s 4
	Common Redshank *Tringa totanus*	m 5
	Common Greenshank *Tringa nebularia*	m 5
	Green Sandpiper *Tringa ochropus*	m 5
	Common Sandpiper *Actitis hypoleucos*	m 5
	Common Black-headed Gull *Larus ridibundus*	m 5
	Brown-headed Gull *Larus brunnicephalus*	m 5
	Lesser Black-backed Gull *Larus fuscus/*	m 5
	Herring Gull *L. argentatus*	m 5
	Common Tern *Sterna aurantia*	m 5
	Hill Pigeon *Columba rupestris*	v
	Snow Pigeon *Columba leuconota*	r 1

+	Speckled Woodpigeon *Columba hodgsonii*	v
	Oriental Turtle Dove *Streptopelia orientalis*	s 4
	Wedge-tailed Green Pigeon *Treron sphenura*	s 5
	Pied Crested Cuckoo *Clamator jacobinus*	v
	Large Hawk-Cuckoo *Hierococcyx sparveriodes*	s 3
	Common Cuckoo *Cuculus canorus*	s 3
	Oriental Cuckoo *Cuculus saturatus*	s 3
	Northern Eagle Owl *Bubo bubo*	v
	Tawny Owl *Strix aluco*	r 4
	Himalayan Swiftlet *Collocalia brevirostris*	s 2
	Pacific Swift *Apus pacificus*	s 2
	Hoopoe *Upupa epops*	s m
	Greater Short-toed Lark *Calandrella brachydactyla*	m 5
	Hume's Short-toed Lark *Calandrella acutirostris*	w m 4
	Oriental Skylark *Alauda gulgula*	s 4
	Horned Lark *Eremophila alpestris*	r b
	Crag Martin *Ptyonoprogne rupestris*	r
	Barn Swallow *Hirundo rustica*	m 5
+	Nepal House-Martin *Delichon nipalensis*	s 4
	Asian House-Martin *Delichon dasypus*	s
	Blyth's Pipit *Anthus godlewskii*	m
	Olive-backed Pipit *Anthus hodgsoni*	s 1
	Rosy Pipit *Anthus roseatus*	s b 2
	Citrine Wagtail *Motacilla citreola*	m 3
	Grey Wagtail *Motacilla cinerea*	s m 3
	White Wagtail *Motacilla alba*	s 2
	Long-tailed Minivet *Pericrocotus ethologus*	s
	White-breasted Dipper *Cinclus cinclus*	? 5
	Brown Dipper *Cinclus pallasii*	r 2
	Northern Wren *Troglodytes troglodytes*	r 2
+	Rufous-breasted Accentor *Prunella strophiata*	s 2
+	Robin Accentor *Prunella rubeculoides*	r 1
	Altai Accentor *Prunella himalayana*	w 1
	Alpine Accentor *Prunella collaris*	r 2
	Bluethroat *Luscinia svecica*	m 5
	White-tailed Rubythroat *Luscinia pectoralis*	s b 3
	Orange-flanked Bush-Robin *Tarsiger cyanurus*	r b 2
+	Golden Bush-Robin *Tarsiger chrysaeus*	s 1 w 5
+	White-browed Bush-Robin *Tarsiger indicus*	s 5
	Black Redstart *Phoenicurus ochruros*	s 3
	Hodgson's Redstart *Phoenicurus hodgsoni*	w 5
+	Blue-fronted Redstart *Phoenicurus frontalis*	r b 1
+	White-throated Redstart *Phoenicurus schisticeps*	r 3
	Güldenstadt's Redstart *Phoenicurus erythrogaster*	r b w 4
	Plumbeous Redstart *Rhyacornis fuliginosus*	s 5
+	White-bellied Redstart *Hodgsonius phoenicuroides*	s 3
+	Grandala *Grandala coelicolor*	r 2
	Common Stonechat *Saxicola torquata*	s? m?
+	White-capped Redstart *Chaimarrornis leucocephalus*	s b 1
	Chestnut-bellied Rock Thrush *Monticola rufiventris*	s
	Blue Whistling Thrush *Myiophoneus caeruleus*	r 2
+	Plain-backed Mountain Thrush *Zoothera mollissima*	s 4

+	Long-tailed Mountain Thrush *Zoothera dixoni*	s 4
+	White-collared Blackbird *Turdus albocinctus*	s 5
	Eurasian Blackbird *Turdus merula*	m 5
	Kessler's Thrush *Turdus kessleri*	v
	Dark-throated Thrush *Turdus ruficollis*	w 4
	Little Forktail *Enicurus scouleri*	s 5
+	Grey-sided Bush Warbler *Cettia brunnifrons*	s b 3
	Blyth's Crowned Warbler *Phylloscopus reguloides*	s 3
	Greenish Warbler *Phylloscopus trochiloides*	s 2
+	Large-billed Leaf Warbler *Phylloscopus magnirostris*	s 2
+	Orange-barred Leaf Warbler *Phylloscopus pulcher*	s b 2
	Grey-faced Leaf Warbler *Phylloscopus maculipennis*	s 5
	Pallas's Leaf Warbler *Phylloscopus proregulus*	s 2
	Yellow-browed Warbler *Phylloscopus inornatus*	s 3
*	Smoky Warbler *Phylloscopus fuligiventer*	s 4
	Tickell's Warbler *Phylloscopus affinis*	s b 2
	Goldcrest *Regulus regulus*	r b 3
	Slaty-blue Flycatcher *Ficedula tricolor*	s 3
	Orange-gorgetted Flycatcher *Ficedula strophiata*	s 3
+	Yellow-bellied Fantail *Rhipidura hypoxantha*	r 3
	Streaked Laughing-thrush *Garrulax lineatus*	r 2
+	Black-faced Laughing-thrush *Garrulax affinis*	r 1
+	White-browed Fulvetta *Alcippe vinipectus*	r b 2
+	Black-capped Sibia *Heterophasia capistrata*	r 4
+	Rufous-vented Yuhina *Yuhina occipitalis*	s 4
+	Grey-crested Tit *Parus dichrous*	r b 2
+	Rufous-vented Black Tit *Parus rubidiventris*	r b 1
	Coal Tit *Parus ater*	r 2
	Green-backed Tit *Parus monticolus*	r 2
	Wallcreeper *Tichodroma muraria*	r? w b
*	Rusty-flanked Treecreeper *Certhia nipalensis*	s 4
	Common Treecreeper *Certhia familiaris*	r? s b
	Mrs Gould's Sunbird *Aethopyga gouldiae*	s 5
+	Fire-tailed Sunbird *Aethopyga ignicauda*	s 5
+	Grey-backed Shrike *Lanius tephronotus*	s 2
	Eurasian Nutcracker *Nucifraga caryocatactes*	r 2
	Alpine Chough *Pyrrhocorax graculus*	r b 1
	Red-billed Chough *Pyrrhocorax pyrrhocorax*	r 1
	Jungle Crow *Corvus macrorhynchos*	r 1
	Common Raven *Corvus corax*	r
	Cinnamon Sparrow *Passer rutilans*	s 5
	Red-necked Snowfinch *Montifrigilla ruficollis*	v
	Yellow-breasted Greenfinch *Carduelis spinoides*	r
	Twite *Carduelis flavirostris*	? 5
	Common Crossbill *Loxia curvirostra*	? 5
	Eurasian Tree Sparrow *Passer montanus*	r 2
	Plain Mountain-Finch *Leucosticte nemoricola*	r b 1
	Brandt's Mountain-Finch *Leucosticte brandti*	r 3
+	Dark-breasted Rosefinch *Carpodacus nipalensis*	s
	Common Rosefinch *Carpodacus erythrinus*	r? s
	Beautiful Rosefinch *Carpodacus pulcherrimus*	r b 1
+	Pink-browed Rosefinch *Carpodacus rhodochrous*	s

+	White-browed Rosefinch *Carpodacus thura*	r 3
+	Crimson-eared Rosefinch *Carpodacus rubicilloides*	? 5
	Spot-crowned Rosefinch *Carpodacus rubicilla*	r b 5
	Red-breasted Rosefinch *Carpodacus puniceus*	r 5
*	Crimson-browed Finch *Propyrrhula subhimachala*	r?
+	Red-headed Bullfinch *Pyrrhula erythrocephala*	r 3
+	Collared Grosbeak *Mycerobas affinis*	? 5
	White-winged Grosbeak *Mycerobas carnipes*	r?1

BARUN VALLEY

	Black Kite *Milvus migrans*
	Lammergeier *Gypaetus barbatus*
	Hen Harrier *Circus cyaneus*
	Northern Goshawk *Accipiter gentilis*
	Crested Goshawk *Accipiter trivirgatus*
	Long-legged Buzzard *Buteo rufinus*
	Upland Buzzard *Buteo hemilasius*
	Black Eagle *Ictinaetus malayensis*
	Common Kestrel *Falco tinnunculus*
E	Red-necked Falcon *Falco chicquera*
	Peregrine *Falco peregrinus*
	Tibetan Snowcock *Tetraogallus tibetanus*
	Common Hill Partridge *Arborophila torqueola*
+	Blood Pheasant *Ithaginis cruentus*
V+	Satyr Tragopan *Tragopan satyra*
+	Himalayan Monal *Lophophorus impejanus*
	Kalij Pheasant *Lophura leucomelana*
	Pintail Snipe *Gallinago stenura*
	Solitary Snipe *Gallinago solitaria*
I*	Wood Snipe *Gallinago nemoricola*
	Snow Pigeon *Columba leuconota*
+	Slaty-headed Parakeet *Psittacula himalayana*
R	Mountain Scops Owl *Otus spilocephalus*
E	Forest Eagle Owl *Bubo nipalensis*
	Collared Owlet *Glaucidium brodiei*
	Brown Hawk Owl *Ninox scutulata*
V	Brown Wood Owl *Strix leptogrammica*
	Tawny Owl *Strix aluco*
	Pacific Swift *Apus pacificus*
	Hoopoe *Upupa epops*
	Great Barbet *Megalaima virens*
	Blue-throated Barbet *Megalaima asiatica*
	Greater Yellow-naped Woodpecker *Picus flavinucha*
	Grey-headed Woodpecker *Picus canus*
+	Darjeeling Pied Woodpecker *Dendrocopos darjellensis*
+	Brown-fronted Pied Woodpecker *Dendrocopos auriceps*
	Crag Martin *Ptyonoprogne rupestris*
	Barn Swallow *Hirundo rustica*
+	Nepal House-Martin *Delichon nipalensis*
	Olive-backed Pipit *Anthus hodgsoni*
	Rosy Pipit *Anthus roseatus*
	Citrine Wagtail *Motacilla citreola*
	Grey Wagtail *Motacilla cinerea*
	White Wagtail *Motacilla alba*
R	Short-billed Minivet *Pericrocotus brevirostris*
	Long-tailed Minivet *Pericrocotus ethologus*
+	Striated Bulbul *Pycnonotus striatus*
	White-cheeked Bulbul *Pycnonotus leucogenys*
	Black Bulbul *Hypsipetes madagascariensis*
	Orange-bellied Leafbird *Chloropsis hardwickii*

Brown Dipper *Cinclus pallasii*
Northern Wren *Troglodytes troglodytes*
Maroon-backed Accentor *Prunella immaculata*
Altai Accentor *Prunella himalayana*
Alpine Accentor *Prunella collaris*
R   White-browed Shortwing *Brachypteryx montana*
Orange-flanked Bush-Robin *Tarsiger cyanurus*
+   Golden Bush-Robin *Tarsiger chrysaeus*
+   White-browed Bush-Robin *Tarsiger indicus*
*   Rufous-breasted Bush-Robin *Tarsiger hyperythrus*
Black Redstart *Phoenicurus ochruros*
+   Blue-fronted Redstart *Phoenicurus frontalis*
Plumbeous Redstart *Rhyacornis fuliginosus*
Common Stonechat *Saxicola torquata*
White-capped Redstart *Chaimarrornis leucocephalus*
Blue Whistling Thrush *Myiophoneus caeruleus*
+   Plain-backed Mountain Thrush *Zoothera mollissima*
+   Long-tailed Mountain-Thrush *Zoothera dixoni*
Scaly Thrush *Zoothera dauma*
*   Long-billed Thrush *Zoothera monticola*
I   Dark-sided Thrush *Zoothera marginata*
*   Pied Ground Thrush *Zoothera wardii*
+   Tickell's Thrush *Turdus unicolor*
+   White-collared Blackbird *Turdus albocinctus*
Chestnut Thrush *Turdus rubrocanus*
Dark-throated Thrush *Turdus ruficollis*
Little Forktail *Enicurus scouleri*
+   Chestnut-headed Tesia *Tesia castaneocoronata*
I   Slaty-bellied Tesia *Tesia olivea*
+   Grey-sided Bush Warbler *Cettia brunnifrons*
Striated Prinia *Prinia criniger*
Common Tailorbird *Orthotomus sutorius*
Golden-spectacled Warbler *Seicercus burkii*
Chestnut-crowned Warbler *Seicercus castaniceps*
+   Grey-hooded Warbler *Seicercus xanthoschistos*
I* Broad-billed Warbler *Abroscopus hodgsoni*
Yellow-bellied Warbler *Abroscopus superciliaris*
+   Black-faced Warbler *Abroscopus schisticeps*
Greenish Warbler *Phylloscopus trochiloides*
+   Large-billed Leaf Warbler *Phylloscopus magnirostris*
+   Orange-barred Leaf Warbler *Phylloscopus pulcher*
Pallas's Leaf Warbler *Phylloscopus proregulus*
Yellow-browed Warbler *Phylloscopus inornatus*
*   Smoky Warbler *Phylloscopus fuliginventer*
Tickell's Warbler *Phylloscopus affinis*
V   Hill Blue Flycatcher *Cyornis banyumas*
R   Ferruginous Flycatcher *Muscicapa ferruginea*
Slaty-blue Flycatcher *Ficedula tricolor*
+   Slaty-backed Flycatcher *Ficedula hodgsonii*
V   White-gorgetted Flycatcher *Ficedula mo’nileger*
Orange-gorgetted Flycatcher *Ficedula strophiata*
+   Yellow-bellied Fantail *Rhipidura hypoxantha*

	Rusty-cheeked Scimitar-Babbler *Pomatorhinus erythrogenys*
I	Coral-billed Scimitar-Babbler *Pomatorhinus ferruginosus*
V+	Slender-billed Scimitar-Babbler *Xiphirhynchus superciliaris*
I*	Spotted Wren-Babbler *Spelaeornis formosus*
E*	Tailed Wren-Babbler *Spelaeornis caudatus*
	Rufous-capped Babbler *Stachyris ruficeps*
+	Black-chinned Babbler *Stachyris pyrrhops*
E	Golden Babbler *Stachyris chrysaea*
	Grey-throated Babbler *Stachyris nigriceps*
R	Black-throated Parrotbill *Paradoxornis nipalensis*
+	Spiny Babbler *Turdoides nipalensis*
+	White-throated Laughing-thrush *Garrulax albogularis*
+	Striated Laughing-thrush *Garrulax striatus*
+	Spotted Laughing-thrush *Garrulax ocellatus*
E+	Blue-winged Laughing-thrush *Garrulax squamatus*
V+	Scaly Laughing-thrush *Garrulax subunicolor*
+	Black-faced Laughing-thrush *Garrulax affinis*
	Chestnut-crowned Laughing-thrush *Garrulax erythrocephalus*
	Red-billed Leiothrix *Leiothrix lutea*
R*	Fire-tailed Myzornis *Myzornis pyrrhoura*
E*	Black-headed Shrike-Babbler *Pteruthius rufiventer*
	White-browed Shrike-Babbler *Pteruthius flaviscapis*
+	Green Shrike-Babbler *Pteruthius xanthochloris*
	Black-eared Shrike-Babbler *Pteruthius melanotis*
E+	Rusty-fronted Barwing *Actinodura egertoni*
	Blue-winged Minla *Minla cyanouroptera*
	Chestnut-tailed Minla *Minla strigula*
+	Red-tailed Minla *Minla ignotincta*
	Rufous-winged Fulvetta *Alcippe castaneceps*
+	White-browed Fulvetta *Alcippe vinipectus*
+	Nepal Fulvetta *Alcippe nipalensis*
+	Black-capped Sibia *Heterophasia capistrata*
+	Whiskered Yuhina *Yuhina flavicollis*
+	Stripe-throated Yuhina *Yuhina gularis*
+	Black-browed Tit *Aegithalos iouschistos*
	Black-throated Tit *Aegithalos concinnus*
	Yellow-browed Tit *Sylviparus modestus*
+	Grey-crested Tit *Parus dichrous*
+	Rufous-vented Black Tit *Parus rubidiventris*
	Coal Tit *Parus ater*
	Green-backed Tit *Parus monticolus*
	Black-lored Tit *Parus xanthogenys*
+	White-tailed Nuthatch *Sitta himalayensis*
	Brown-throated Treecreeper *Certhia discolor*
*	Rusty-flanked Treecreeper *Certhia nipalensis*
	Green-tailed Sunbird *Aethopyga nipalensis*
+	Fire-tailed Sunbird *Aethopyga ignicauda*
R*	Yellow-bellied Flowerpecker *Dicaeum melanoxanthum*
+	Grey-backed Shrike *Lanius tephronotus*
	Ashy Drongo *Dicrurus leucophaeus*
	Lesser Racket-tailed Drongo *Dicrurus remifer*
+	Yellow-billed Blue Magpie *Urocissa flavirostris*

Grey Treepie *Dendrocitta formosae*
Eurasian Nutcracker *Nucifraga caryocatactes*
Red-billed Chough *Pyrrhocorax pyrrhocorax*
Jungle Crow *Corvus macrorhynchos*
Common Raven *Corvus corax*
Plain Mountain-Finch *Leucosticte nemoricola*
Brandt's Mountain-Finch *Leucosticte brandti*
+ Dark-breasted Rosefinch *Carpodacus nipalensis*
+ Pink-browed Rosefinch *Carpodacus rhodochrous*
* Spot-winged Rosefinch *Carpodacus rhodopeplus*
+ White-browed Rosefinch *Carpodacus thura*
+ Gold-naped Finch *Pyrrhoplectes epauletta*
Brown Bullfinch *Pyrrhula nipalensis*
+ Red-headed Bullfinch *Pyrrhula erythrocephala*
White-winged Grosbeak *Mycerobas carnipes*

RARA LAKE NATIONAL PARK

Little Grebe *Tachybaptus ruficollis*		r?
Great Crested Grebe *Podiceps cristatus*		r? w
Black-necked Grebe *Podiceps nigricollis*		r? w
Great Cormorant *Phalacrocorax carbo*		s?
Eurasian Bittern *Botauris stellaris*		m 5
Great Egret *Egretta alba*		m 5
Grey Heron *Ardea cinerea*		m 5
Greylag Goose *Anser anser*		m
Bar-headed Goose *Anser indicus*		m
Ruddy Shelduck *Tadorna ferruginea*		m?
Eurasian Wigeon *Anas penelope*		w m
Gadwall *Anas strepera*		m
Common Teal *Anas crecca*		w m
Mallard *Anas platyrhynchos*		w m
Northern Pintail *Anas acuta*		m
Northern Shoveler *Anas clypeata*		m
Red-crested Pochard *Netta rufina*		w m
Common Pochard *Aythya ferina*		m
Ferruginous Duck *Aythya nyroca*		m
Tufted Duck *Aythya fuligula*		w m
Common Goldeneye *Bucephala clangula*		w m
Goosander *Mergus merganser*		w
Black Kite *Milvus migrans*		s b?
Lammergeier *Gypaetus barbatus*		r
Himalayan Griffon Vulture *Gyps himalayensis*		r
Hen Harrier *Circus cyaneus*		w m
Pallid Harier *Circus macrourus*		m
Northern Goshawk *Accipiter gentilis*		r?
Northern Sparrowhawk *Accipiter nisus*		s?
Common Buzzard *Buteo buteo*		w m
Long-legged Buzzard *Buteo rufinus*		w m
Black Eagle *Ictinaetus malayensis*		r
Osprey *Pandion haliaetus*		m
Common Kestrel *Falco tinnunculus*		r?
Eurasian Hobby *Falco subbuteo*		m?
Himalayan Snowcock *Tetraogallus himalayensis*		r
Chukar Partridge *Alectoris chukar*		r?
+	Blood Pheasant *Ithaginis cruentus*	r
+	Himalayan Monal *Lophophorus impejanus*	r
	Kalij Pheasant *Lophura leucomelana*	s?
I*	Cheer Pheasant *Catreus wallichii*	r
	Common Moorhen *Gallinula chloropus*	m?
	Common Coot *Fulica atra*	w m
	Pheasant-tailed Jacana *Hydrophasianus chirurgus*	m
	Lesser Sand Plover *Charadrius mongolus*	m
	Little Stint *Calidris minuta*	m
	Temminck's Stint *Calidris temminckii*	m
	Common Snipe *Gallinago gallinago*	m
	Common Redshank *Tringa totanus*	m
	Common Greenshank *Tringa nebularia*	m

	Green Sandpiper *Tringa ochropus*	m
	Wood Sandpiper *Tringa glareola*	m
	Common Sandpiper *Actitis hypoleucos*	m
	Red-necked Phalarope *Phalaropus lobatus*	m
	Great Black-headed Gull *Larus ichthyaetus*	m
	Common Black-headed Gull *Larus ridibundus*	m
	Brown-headed Gull *Larus brunnicephalus*	m
	Herring Gull *Larus argentatus*/	
	Lesser Black-backed Gull *Larus fuscus*	m
	Gull-billed Tern *Gelochelidon nilotica*	m
	Rock Pigeon *Columba livia*	s?
	Hill Pigeon *Columba rupestris*	w
	Snow Pigeon *Columba leuconota*	r?
	Oriental Turtle Dove *Streptopelia orientalis*	s
	Common Cuckoo *Cuculus canorus*	s
	White-throated Needletail *Hirundapus caudacutus*	?
	Crested Kingfisher *Ceryle lugubris*	s?
	Hoopoe *Upupa epops*	s m
	Scaly-bellied Green Woodpecker *Picus squamatus*	r
+	Himalayan Pied Woodpecker *Dendrocopos himalayensis*	r
	Oriental Skylark *Alauda gulgula*	s?
	Olive-backed Pipit *Anthus hodgsoni*	s
	Red-throated Pipit *Anthus cervinus*	m 5
	Yellow Wagtail *Motacilla flava*	m
	Citrine Wagtail *Motacilla citreola*	m
	Grey Wagtail *Motacilla cinerea*	s
	White Wagtail *Motacilla alba*	s m?
	Long-tailed Minivet *Pericrocotus ethologus*	s b
	White-cheeked Bulbul *Pycnonotus leucogenys*	r
	Brown Dipper *Cinclus pallasii*	r
	Northern Wren *Troglodytes troglodytes*	r
+	Rufous-breasted Accentor *Prunella strophiata*	r?
	Brown Accentor *Prunella fulvescens*	?
	Black-throated Accentor *Prunella atrogularis*	w
	Altai Accentor *Prunella himalayana*	w
	Alpine Accentor *Prunella collaris*	w?
	Orange-flanked Bush-Robin *Tarsiger cyanurus*	s?
	Rufous-backed Redstart *Phoenicurus erythronotus*	w
	Blue-capped Redstart *Phoenicurus caeruleocephalus*	s? b
+	Blue-fronted Redstart *Phoenicurus frontalis*	r?
	Plumbeous Redstart *Rhyacornis fuliginosus*	s?
	Grey Bushchat *Saxicola ferrea*	s?
	Desert Wheatear *Oenanthe deserti*	m?
	White-capped Redstart *Chaimarrornis leucocephalus*	s
	Blue Whistling Thrush *Myiophoneus caeruleus*	s
+	Long-tailed Mountain Thrush *Zoothera dixoni*	s
+	Tickell's Thrush *Turdus unicolor*	m?
+	White-collared Blackbird *Turdus albocinctus*	r?
	Dark-throated Thrush *Turdus ruficollis*	w
	Mistle Thrush *Turdus viscivorus*	r?
	Little Forktail *Enicurus scouleri*	s
	Spotted Forktail *Enicurus maculatus*	s?

+	Orange-barred Leaf Warbler *Phylloscopus pulcher*	r?
+	Large-billed Leaf Warbler *Phylloscopus magnirostris*	s
	Goldcrest *Regulus regulus*	r?
	Asian Sooty Flycatcher *Muscicapa sibirica*	s
+	Rufous-tailed Flycatcher *Muscicapa ruficauda*	s
	Slaty-blue Flycatcher *Ficedula tricolor*	s
+	Ultramarine Flycatcher *Ficedula superciliaris*	s
	Orange-gorgetted Flycatcher *Ficedula strophiata*	s
+	Yellow-bellied Fantail *Rhipidura hypoxantha*	s
+	Variegated Laughing-thrush *Garrulax variegatus*	r
+	Spotted Laughing-thrush *Garrulax ocellatus*	r
	Streaked Laughing-thrush *Garrulax lineatus*	s?
	Chestnut-crowned Laughing-thrush *Garrulax erythrocephalus*	s
	Chestnut-tailed Minla *Minla strigula*	s
+	White-browed Fulvetta *Alcippe vinipectus*	r?
+	Stripe-throated Yuhina *Yuhina gularis*	r?
*	White-throated Tit *Aegithalos niveogularis*	r?
	Black-throated Tit *Aegithalos concinnus*	r?
+	Grey-crested Tit *Parus dichrous*	r
	Rufous-naped Tit *Parus rufonuchalis*	r?
+	Rufous-vented Black Tit *Parus rubidiventris*	r?
+	Spot-winged Black Tit *Parus melanolophus*	r?
	Great Tit *Parus major*	s?
	Green-backed Tit *Parus monticolus*	s?
+	White-cheeked Nuthatch *Sitta leucopsis*	r?
+	Kashmir Nuthatch *Sitta cashmirensis*	r?
	Bar-tailed Treecreeper *Certhia himalayana*	r?
*	Rusty-flanked Treecreeper *Certhia nipalensis*	r
	Common Treecreeper *Certhia familiaris*	r?
	Long-tailed Shrike *Lanius schach*	s?
+	Grey-backed Shrike *Lanius tephronotus*	s?
+	Lanceolated Jay *Garrulus lanceolatus*	s?
+	Yellow-billed Blue Magpie *Urocissa flavirostris*	s?
	Red-billed Blue Magpie *Urocissa erythrorhyncha*	m
	Eurasian Nutcracker *Nucifraga caryocatactes*	r?
	Red-billed Chough *Pyrrhocorax pyrrhocorax*	r?
	Jungle Crow *Corvus macrorhynchos*	r
	Common Raven *Corvus corax*	r
	Eurasian Tree Sparrow *Passer montanus*	r
	Common Chaffinch *Fringilla coelebs*	w
	Brambling *Fringilla montifringilla*	w
	Red-fronted Serin *Serinus pusillus*	r?
	Yellow-breasted Greenfinch *Carduelis spinoides*	s
	Eurasian Goldfinch *Carduelis carduelis*	s
	Plain Mountain-Finch *Leucosticte nemoricola*	r
	Common Rosefinch *Carpodacus erythrinus*	s
	Beautiful Rosefinch *Carpodacus pulcherrimus*	r
	Pink-browed Rosefinch *Carpodacus rhodochrous*	r?
+	Crimson-eared Rosefinch *Carpodacus rubicilloides*	w?
	Red-headed Bullfinch *Pyrrhula erythrocephala*	r?
	Collared Grosbeak *Mycerobas affinis*	r?
	Rock Bunting *Emberiza cia*	r

SHEY-PHOKSUNDO NATIONAL PARK

	Black Kite *Milvus migrans*	s
	Lammergeier *Gypaetus barbatus*	r
	Himalayan Griffon Vulture *Gyps himalayensis*	r
	Hen Harrier *Circus cyaneus*	w m
	Northern Goshawk *Accipiter gentilis*	r?
	Northern Sparrowhawk *Accipiter nisus*	s
	Common Buzzard *Buteo buteo*	?
	Golden Eagle *Aquila chrysaetos*	r
	Common Kestrel *Falco tinnunculus*	r? s? m?
	Amur Falcon *Falco amurensis*	m
	Merlin *Falco columbarius*	w
+	Snow Partridge *Lerwa lerwa*	r
	Tibetan Snowcock *Tetraogallus tibetanus*	r
	Himalayan Snowcock *Tetraogallus himalayensis*	r
	Chukar Partridge *Alectoris chukar*	r
	Tibetan Partridge *Perdix hodgsoniae*	r
+	Himalayan Monal *Lophophorus impejanus*	r
	Rock Pigeon *Columba livia*	r
	Hill Pigeon *Columba rupestris*	r
	Snow Pigeon *Columba leuconota*	r
	Oriental Turtle Dove *Streptopelia orientalis*	s
	Common Cuckoo *Cuculus canorus*	s
	Northern Little Owl *Athene noctua*	r
	Tawny Owl *Strix aluco*	r
	Short-eared Owl *Asio flammeus*	m
	Himalayan Swiftlet *Collocalia brevirostris*	s
	Alpine Swift *Apus melba*	s
	Hoopoe *Upupa epops*	s? m?
	Greater Short-toed Lark *Calandrella brachydactyla*	m?
	Hume's Short-toed Lark *Calandrella acutirostris*	s
	Oriental Skylark *Alauda gulgula*	s
	Horned Lark *Eremophila alpestris*	r
	Crag Martin *Ptyonoprogne rupestris*	s
	Rosy Pipit *Anthus roseatus*	s
	White Wagtail *Motacilla alba*	s m?
	Long-tailed Minivet *Pericrocotus ethologus*	s?
	White-breasted Dipper *Cinclus cinclus*	r
	Brown Dipper *Cinclus pallasii*	r
	Northern Wren *Troglodytes troglodytes*	r
	Brown Accentor *Prunella fulvescens*	r
	Black-throated Accentor *Prunella atrogularis*	w
+	Robin Accentor *Prunella rubeculoides*	r
	Alpine Accentor *Prunella collaris*	r
	White-tailed Rubythroat *Luscinia pectoralis*	s
	Orange-flanked Bush-Robin *Tarsiger cyanurus*	r?
+	Golden Bush-Robin *Tarsiger chrysaeus*	r?
	Blue-capped Redstart *Phoenicurus caeruleocephalus*	r? s?
	Black Redstart *Phoenicurus ochruros*	s
+	Blue-fronted Redstart *Phoenicurus frontalis*	r?
+	White-throated Redstart *Phoenicurus schisticeps*	r?

	Güldenstadt's Redstart *Phoenicurus erythrogaster*	r?
	Plumbeous Redstart *Rhyacornis fuliginosus*	r? s?
+	White-bellied Redstart *Hodgsonius phoenicuroides*	s
	Desert Wheatear *Oenanthe deserti*	s
	Common Stonechat *Saxicola torquata*	s
	Blue Rock-Thrush *Monticola solitarius*	s
	Blue Whistling Thrush *Myiophoneus caeruleus*	r? s?
	Dark-throated Thrush *Turdus ruficollis*	w
	Mistle Thrush *Turdus viscivorus*	r
	Little Forktail *Enicurus scouleri*	s
+	Grey-sided Bush Warbler *Cettia brunnifrons*	s?
	Striated Prinia *Prinia criniger*	s?
	Golden-spectacled Warbler *Seicercus burkii*	s?
	Greenish Warbler *Phylloscopus trochiloides*	s
	Pallas's Leaf Warbler *Phylloscopus proregulus*	s?
	Yellow-browed Warbler *Phylloscopus inornatus*	s m?
	Tickell's Warbler *Phylloscopus affinis*	s
	Goldcrest *Regulus regulus*	r
	Stoliczka's Tit-Warbler *Leptopoecile sophiae*	r
+	Yellow-bellied Fantail *Rhipidura hypoxantha*	s
+	Variegated Laughing-thrush *Garrulax variegatus*	r?
	Streaked Laughing-thrush *Garrulax lineatus*	r?
+	White-browed Fulvetta *Alcippe vinipectus*	r?
*	White-throated Tit *Aegithalos niveogularis*	r
	Rufous-naped Black Tit *Parus rufonuchalis*	r
+	Rufous-vented Black Tit *Parus rubidiventris*	r
+	Spot-winged Black Tit *Parus melanolophus*	r?
+	White-cheeked Nuthatch *Sitta leucopsis*	r?
+	Kashmir Nuthatch *Sitta cashmirensis*	r
	Bar-tailed Treecreeper *Certhia himalayana*	r?
	Mrs Gould's Sunbird *Aethopyga gouldiae*	r? s?
+	Grey-backed Shrike *Lanius tephronotus*	s
	Spangled Drongo *Dicrurus hottentottus*	v
	Hume's Ground Jay *Pseudopodoces humilis*	r
	Eurasian Nutcracker *Nucifraga caryocatactes*	r
	Alpine Chough *Pyrrhocorax graculus*	r
	Red-billed Chough *Pyrrhocorax pyrrhocorax*	r
	Jungle Crow *Corvus macrorhynchos*	r
	Common Raven *Corvus corax*	r
	House Sparrow *Passer domesticus*	s?
	Cinnamon Sparrow *Passer rutilans*	r? s?
	Eurasian Tree Sparrow *Passer montanus*	r?
	Tibetan Snowfinch *Montifringilla adamsi*	r
	Red-fronted Serin *Serinus pusillus*	r
	Twite *Carduelis flavirostris*	r?
	Plain Mountain-Finch *Leucosticte nemoricola*	r
	Brandt's Mountain-Finch *Leucosticte brandti*	r
	Common Rosefinch *Carpodacus erythrinus*	s
	Beautiful Rosefinch *Carpodacus pulcherrimus*	r?
+	Crimson-eared Rosefinch *Carpodacus rubicilloides*	s
	Spot-crowned Rosefinch *Carpodacus rubicilla*	r
	Red-breasted Rosefinch *Carpodacus puniceus*	r

White-winged Grosbeak *Mycerobas carnipes* r
Pine Bunting *Emberiza leucocephalos* w
Rock Bunting *Emberiza cia* r

KHAPTAD NATIONAL PARK

	Crested Honey Buzzard *Pernis ptilorhyncus*	m 2
	Black Kite *Milvus migrans*	r? 2
	Egyptian Vulture *Neophron percnopterus*	s m 2
	Lammergeier *Gypaetus barbatus*	r 1
	Oriental White-backed Vulture *Gyps bengalensis*	s m 2
	Himalayan Griffon Vulture *Gyps himalayensis*	r 1
	Eurasian Griffon Vulture *Gyps fulvus*	r? s? m 1
	Red-headed Vulture *Sarcogyps calvus*	r 2
	Crested Serpent Eagle *Spilornis cheela*	s 2
	Eurasian Marsh Harrier *Circus aeruginosus*	m 5
	Hen Harrier *Circus cyaneus*	w? m?
	Northern Goshawk *Accipiter gentilis*	r? 5
	Northern Sparrowhawk *Accipiter nisus*	s? 4
	Shikra *Accipiter badius*	r 5
	Common Buzzard *Buteo buteo*	r? w? m? 2
	Upland Buzzard *Buteo hemilasius*	r? w? m? 2
	Black Eagle *Ictinaetus malayensis*	r 3
	Steppe Eagle *Aquila rapax nipalensis*	w? m? 3
	Booted Eagle *Hieraaetus pennatus*	m 5
	Mountain Hawk-Eagle *Spizaetus nipalensis*	r 3
	Common Kestrel *Falco tinnunculus*	r 2 w? m?
	Eurasian Hobby *Falco subbuteo*	r 3 w? m?
	Peregrine *Falco peregrinus*	r 4
	Common Hill Partridge *Arborophila torqueola*	r 1
V+	Satyr Tragopan *Tragopan satyra*	r 4
	Koklass Pheasant *Pucrasia macrolopha*	r 1
+	Himalayan Monal *Lophophorus impejanus*	r 3
	Kalij Pheasant *Lophura leucomelana*	r 3
	Common Coot *Fulica atra*	m 5
	Solitary Snipe *Gallinago solitaria*	r? w? m? 4
	Eurasian Woodcock *Scolopax rusticola*	s 2
	Green Sandpiper *Tringa ochropus*	m
	Common Sandpiper *Actitis hypoleucos*	m
	Rock Pigeon *Columba livia*	?
	Snow Pigeon *Columba leuconota*	w
+	Speckled Woodpigeon *Columba hodgsonii*	r 2
	Oriental Turtle Dove *Streptopelia orientalis*	r 1
	Spotted Dove *Streptopelia chinensis*	r 2
	Emerald Dove *Chalcophaps indica*	r
	Wedge-tailed Green Pigeon *Treron sphenura*	r 2
	Large Hawk-Cuckoo *Hierococcyx sparverioides*	s 2
	Indian Cuckoo *Cuculus micropterus*	s 2
	Common Cuckoo *Cuculus canorus*	s 2
	Oriental Cuckoo *Cuculus saturatus*	s 2
	Lesser Cuckoo *Cuculus poliocephalus*	s 2
	Drongo Cuckoo *Surniculus lugubris*	s
	Collared Scops Owl *Otus lempiji*	r
R	Mountain Scops Owl *Otus spilocephalus*	r 3
	Collared Owlet *Glaucidium brodiei*	r 2
	Asian Barred Owlet *Glaucidium cuculoides*	r 4

V	Brown Wood Owl *Strix leptogrammica*	r
	Tawny Owl *Strix aluco*	r 4
	Jungle Nightjar *Caprimulgus indicus*	s 2
	Himalayan Swiftlet *Collocalia brevirostris*	r 2
	White-throated Needletail *Hirundapus caudacutus*	s? m? 3
	Pacific Swift *Apus pacificus*	? 3
	Alpine Swift *Apus melba*	s? m? 4
	Common Kingfisher *Alcedo atthis*	s 5
	Crested Kingfisher *Ceryle lugubris*	r 3
	Hoopoe *Upupa epops*	m
	Great Barbet *Megalaima virens*	r 1
	Blue-throated Barbet *Megalaima asiatica*	r 3
	Speckled Piculet *Picumnus innominatus*	r
	Rufous Woodpecker *Celeus brachyurus*	r
	Greater Yellow-naped Woodpecker *Picus flavinucha*	r
	Grey-headed Woodpecker *Picus canus*	r 2
	Scaly-bellied Green Woodpecker *Picus squamatus*	r 3
+	Himalayan Pied Woodpecker *Dendrocopos himalayensis*	r 2
	Rufous-bellied Pied Woodpecker *Dendrocopos hyperythrus*	r 2
+	Brown-fronted Pied Woodpecker *Dendrocopos auriceps*	r 2
	Hume's Short-toed Lark *Calandrella acutirostris*	m 5
	Oriental Skylark *Alauda gulgula*	s? 4
	Barn Swallow *Hirundo rustica*	m 5
	Red-rumped Swallow *Hirundo daurica*	m 4
+	Nepal House-Martin *Delichon nipalensis*	m 3
	Olive-backed Pipit *Anthus hodgsoni*	r 1
	Tree Pipit *Anthus trivialis*	m 5
	Rosy Pipit *Anthus roseatus*	m 2
	Upland Pipit *Anthus sylvanus*	r
	Citrine Wagtail *Motacilla citreola*	m 5
	Grey Wagtail *Motacilla cinerea*	s 2
	White Wagtail *Motacilla alba*	m
	Bar-winged Flycatcher-shrike *Hemipus picatus*	r
	Black-winged Cuckoo-shrike *Coracina melaschistos*	s 3
	Scarlet Minivet *Pericrocotus flammeus*	r
	Long-tailed Minivet *Pericrocotus ethologus*	r 1
	White-cheeked Bulbul *Pycnonotus leucogenys*	r
	Red-vented Bulbul *Pycnonotus cafer*	r
	Ashy Bulbul *Hypsipetes flavalus*	r
	Black Bulbul *Hypsipetes madagascariensis*	r 1
	Orange-bellied Leafbird *Chloropsis hardwickii*	r
	Brown Dipper *Cinclus pallasii*	r 4
+	Rufous-breasted Accentor *Prunella strophiata*	w? m? 2
	Altai Accentor *Prunella himalayana*	w
	Bluethroat *Luscinia svecica*	m
	White-tailed Rubythroat *Luscinia pectoralis*	m?
+	Indian Blue Robin *Luscinia brunnea*	s 1
	Orange-flanked Bush-Robin *Tarsiger cyanurus*	r 2
	White-browed Bush-Robin *Tarsiger indicus*	? 5
	Black Redstart *Phoenicuros ochrurus*	m
+	Blue-fronted Redstart *Phoenicurus frontalis*	r 4 w? m?1?
	Plumbeous Redstart *Rhyacornis fuliginosus*	r 2

+	White-bellied Redstart *Hodgsonius phoenicuroides*	s
	Common Stonechat *Saxicola torquata*	m
	Grey Bushchat *Saxicola ferrea*	r 1
	Desert Wheatear *Oenanthe deserti*	m
	White-capped Redstart *Chaimarrornis leucocephalus*	r? s? m 2
+	Blue-capped Rock-Thrush *Monticola cinclorhyncha*	s
	Chestnut-bellied Rock-Thrush *Monticola rufiventris*	r 2
	Blue Whistling Thrush *Myiophoneus caeruleus*	r 3
	Scaly Thrush *Zoothera dauma*	s 3
*	Pied Ground Thrush *Zoothera wardii*	s 4
+	Tickell's Thrush *Turdus unicolor*	s 3
+	White-collared Blackbird *Turdus albocinctus*	r 1
+	Grey-winged Blackbird *Turdus boulboul*	r 2
	Dark-throated Thrush *Turdus ruficollis*	w 2
	Mistle Thrush *Turdus viscivorus*	r 1
	Slaty-backed Forktail *Enicurus schistaceus*	r 4
	Spotted Forktail *Enicurus maculatus*	r 4
+	Chestnut-headed Tesia *Tesia castaneocoronata*	r? s? 2
	Grey-bellied Tesia *Tesia cyaniventer*	r? s? 4
+	Aberrant Bush Warbler *Cettia flavolivacea*	r? s? 2
I	Yellow-bellied Bush Warbler *Cettia acanthizoides*	r? s? 1
+	Grey-sided Bush Warbler *Cettia brunnifrons*	r? s? 2
	Striated Prinia *Prinia criniger*	r
	Blyth's Reed Warbler *Acrocephalus dumetorum*	m 3
	Golden-spectacled Warbler *Seicercus burkii*	r 2
+	Grey-hooded Warbler *Seicercus xanthoschistos*	r 1
+	Black-faced Warbler *Abroscopus schisticeps*	r 3
	Blyth's Crowned Warbler *Phylloscopus reguloides*	s m 1
	Slender-billed Warbler *Phylloscopus tytleri*	s? m? 3
	Green Warbler *Phylloscopus nitidus*	m 5
	Greenish Warbler *Phylloscopus trochiloides*	m 4
+	Large-billed Leaf Warbler *Phylloscopus magnirostris*	s? m? 3
+	Orange-barred Leaf Warbler *Phylloscopus pulcher*	m 2
	Grey-faced Leaf Warbler *Phylloscopus maculipennis*	r 3
	Pallas's Leaf Warbler *Phylloscopus proregulus*	m 3
	Yellow-browed Warbler *Phylloscopus inornatus*	r m 2
*	Smoky Warbler *Phylloscopus fuligiventer*	m 4
	Tickell's Warbler *Phylloscopus affinis*	m
	Goldcrest *Regulus regulus*	r 4
	Small Niltava *Niltava macgrigoriae*	s 2
+	Rufous-bellied Niltava *Niltava sundara*	r? s? 4
	Verditer Flycatcher *Muscicapa thalassina*	s 1
	Asian Sooty Flycatcher *Muscicapa sibirica*	s 1
+	Rufous-tailed Flycatcher *Muscicapa ruficauda*	s? m? 5
	Slaty-blue Flycatcher *Ficedula tricolor*	r 1
+	Ultramarine Flycatcher *Ficedula superciliaris*	s 1
	Orange-gorgetted Flycatcher *Ficedula strophiata*	r? s? 1
	Grey-headed Flycatcher *Culicicapa ceylonensis*	r? s? 1
+	Yellow-bellied Fantail *Rhipidura hypoxantha*	r? s? 1
	Streak-breasted Scimitar-Babbler *Pomatorhinus ruficollis*	r 3
+	Greater Scaly-breasted Wren-Babbler *Pnoepyga albiventer*	r? s? 4
	Lesser Scaly-breasted Wren-Babbler *Pnoepyga pusilla*	r? s? 4

+	Black-chinned Babbler *Stachyris pyrrhops*	r 3
V*	Great Parrotbill *Conostoma aemodium*	r 4
R	Black-throated Parrotbill *Paradoxornis nipalensis*	r 1
+	White-throated Laughing-thrush *Garrulax albogularis*	r 1
	White-crested Laughing-thrush *Garrulax leucolophus*	r 2
+	Striated Laughing-thrush *Garrulax striatus*	r 2
+	Variegated Laughing-thrush *Garrulax variegatus*	r 2
+	Spotted Laughing-thrush *Garrulax ocellatus*	r 2
	Streaked Laughing-thrush *Garrulax lineatus*	r 2
	Chestnut-crowned Laughing-thrush *Garrulax erythrocephalus*	r 2
+	Green Shrike-Babbler *Pteruthius xanthochloris*	r 3
*	Hoary Barwing *Actinodura nipalensis*	r 2
	Blue-winged Minla *Minla cyanouroptera*	r 3
	Chestnut-tailed Minla *Minla strigula*	r 1
+	White-browed Fulvetta *Alcippe vinipectus*	r 1
+	Black-capped Sibia *Heterophasia capistrata*	r 1
+	Whiskered Yuhina *Yuhina flavicollis*	r 1
+	Stripe-throated Yuhina *Yuhina gularis*	r 1
E	Black-chinned Yuhina *Yuhina nigrimenta*	r 2
	Black-throated Tit *Aegithalos concinnus*	r 1
	Yellow-browed Tit *Sylviparus modestus*	r 3
+	Grey-crested Tit *Parus dichrous*	r 2
+	Rufous-vented Black Tit *Parus rubidiventris*	r? w?
+	Spot-winged Black Tit *Parus melanolophus*	r 1
	Green-backed Tit *Parus monticolus*	r 1
	Black-lored Tit *Parus xanthogenys*	r 1
+	White-tailed Nuthatch *Sitta himalayensis*	r 2
	Bar-tailed Treecreeper *Certhia himalayana*	r 3
*	Rusty-flanked Treecreeper *Certhia nipalensis*	r 3
	Common Treecreeper *Certhia familiaris*	r 4
+	Fire-capped Tit *Cephalopyrus flammiceps*	r? 3
	Mrs Gould's Sunbird *Aethopyga gouldiae*	r 2
	Green-tailed Sunbird *Aethopyga nipalensis*	r 1
	Black-throated Sunbird *Aethopyga saturata*	r 2
+	Fire-tailed Sunbird *Aethopyga ignicauda*	s? 2
	Buff-bellied Flowerpecker *Dicaeum ignipectus*	r 2
	Oriental White Eye *Zosterops palpebrosa*	s?
	Maroon Oriole *Oriolus traillii*	r? s? 2
	Long-tailed Shrike *Lanius schach*	s 3
	Ashy Drongo *Dicrurus leucophaeus*	s 1
	Bronzed Drongo *Dicrurus aeneus*	s
	Lesser Racket-tailed Drongo *Dicrurus remifer*	s
	Eurasian Jay *Garrulus glandarius*	r 2
+	Lanceolated Jay *Garrulus lanceolatus*	r 2
+	Yellow-billed Blue Magpie *Urocissa flavirostris*	r 1
	Red-billed Blue Magpie *Urocissa erythrorhyncha*	r
	Green Magpie *Cissa chinensis*	r
	Grey Treepie *Dendrocitta formosae*	r? s? 2
	Eurasian Nutcracker *Nucifraga caryocatactes*	r 2
	Jungle Crow *Corvus macrorhynchos*	r 1
	Common Mynah *Acridotheres tristis*	r? s?
	Jungle Mynah *Acridotheres fuscus*	r? s?

	Cinnamon Sparrow *Passer rutilans*	r
	Red-fronted Serin *Serinus pusillus*	w? m?
	Yellow-breasted Greenfinch *Carduelis spinoides*	s? m?
	Common Crossbill *Loxia curvirostra*	r 2
+	Dark-breasted Rosefinch *Carpodacus nipalensis*	r? w? m? 4
	Common Rosefinch *Carpodacus erythrinus*	w? m? 2
	Beautiful Rosefinch *Carpodacus pulcherrimus*	w? m?
+	Pink-browed Rosefinch *Carpodacus rhodochrous*	r? w? m? 2
*	Spot-winged Rosefinch *Carpodacus rhodopeplus*	r? w? m? 3
+	Red-headed Bullfinch *Pyrrhula erythrocephala*	r? 4
+	Collared Grosbeak *Mycerobas affinis*	r 2
	Rock Bunting *Emberiza cia*	r ?

ROYAL SUKLA PHANTA WILDLIFE RESERVE

	Little Grebe *Tachybaptus ruficollis*	r 5
	Great Cormorant *Phalacrocorax carbo*	r w? 1
	Little Cormorant *Phalacrocorax niger*	r 3
	Darter *Anhinga melanogaster*	r 2
	Cinnamon Bittern *Ixobrychus cinnamomeus*	r 5
V	Black Bittern *Dupetor flavicollis*	r 5
	Black-crowned Night Heron *Nycticorax nycticorax*	w 3
	Green-backed Heron *Butorides striatus*	r 3
	Indian Pond Heron *Ardeola grayii*	r 1
	Cattle Egret *Bubulcus ibis*	r 1
	Little Egret *Egretta garzetta*	r 3
	Intermediate Egret *Egretta intermedia*	r 3
	Great Egret *Egretta alba*	r 3
	Grey Heron *Ardea cinerea*	w 3
	Purple Heron *Ardea purpurea*	r 3
	Painted Stork *Mycteria leucocephala*	r? w? s? 3
	Asian Openbill Stork *Anastomus oscitans*	r m w? 3
	Woolly-necked Stork *Ciconia episcopus*	w? s? 2
	Black-necked Stork *Xenorhynchus asiaticus*	r 3
	Lesser Adjutant Stork *Leptoptilos javanicus*	r 3
	Red-naped Ibis *Pseudibis papillosa*	r? 3
	Lesser Whistling Duck *Dendrocygna javanica*	w m 3
	Bar-headed Goose *Anser indicus*	m
	Ruddy Shelduck *Tadorna ferruginea*	w r? 2
	Comb Duck *Sarkidiornis melanotos*	r 3
	Cotton Pygmy Goose *Nettapus coromandelianus*	r 2
	Eurasian Wigeon *Anas penelope*	w m 3
	Falcated Duck *Anas falcata*	w m 5
	Gadwall *Anas strepera*	w m 3
	Mallard *Anas platyrhynchos*	w m 3
	Spotbill *Anas poecilorhyncha*	w m 3
	Northern Pintail *Anas acuta*	w m 2
	Garganey *Anas querquedula*	w? m 3
	Red-crested Pochard *Netta rufina*	w m 2
	Common Pochard *Aythya ferina*	w m 3
	Ferruginous Duck *Aythya nyroca*	w m 3
	Crested Honey Buzzard *Pernis ptilorhyncus*	r? 3
	Black-shouldered Kite *Elanus caeruleus*	r 3
	Black Kite *Milvus migrans*	r m 2
	Brahminy Kite *Haliastur indus*	r 3
E	Pallas's Fish Eagle *Haliaeetus leucoryphus*	r?
E	Lesser Fishing Eagle *Ichthyophaga nana*	r 5
E	Grey-headed Fishing Eagle *Ichthyophaga ichthyaetus*	r 4
	Egyptian Vulture *Neophron percnopterus*	w r? 3
	Oriental White-backed Vulture *Gyps bengalensis*	r 1
	Long-billed Vulture *Gyps indicus*	r? m 2
	Eurasian Griffon Vulture *Gyps fulvus*	w? m? 5
	Red-headed Vulture *Sarcogyps calvus*	r? m? 3
	Crested Serpent Eagle *Spilornis cheela*	r? w 2
	Marsh Harrier *Circus aeruginosus*	w 3

	Hen Harrier *Circus cyaneus*	w? m 3
	Pied Harrier *Circus melanoleucus*	w 3
	Shikra *Accipiter badius*	r 3
	White-eyed Buzzard *Butastur teesa*	r 2
R	Tawny Eagle *Aquila rapax vindhiana*	r 2
	Booted Eagle *Hieraaetus pennatus*	w? m
V	Changeable Hawk-Eagle *Spizaetus cirrhatus*	r 3
	Osprey *Pandion haliaetus*	w? m 5
	Black Francolin *Francolinus francolinus*	r 1
V*	Swamp Francolin *Francolinus gularis*	r 4
	Red Junglefowl *Gallus gallus*	r 1
	Blue Peafowl *Pavo cristatus*	r 1
	Barred Buttonquail *Turnix suscitator*	r
	Ruddy-breasted Crake *Porzana fusca*	r? w 5
	White-breasted Waterhen *Amaurornis phoenicurus*	r 2
	Common Moorhen *Gallinula chloropus*	r 1
	Purple Gallinule *Porphyrio porphyrio*	r 1
	Common Coot *Fulica atra*	r? w 2
	Sarus Crane *Grus antigone*	r 3
	Demoiselle Crane *Anthropoides virgo*	m 5
E*	Bengal Florican *Houbaropsis bengalensis*	r? 3
	Pheasant-tailed Jacana *Hydrophasianus chirurgus*	r 3
	Bronze-winged Jacana *Metopidius indicus*	r 2
	Painted Snipe *Rostratula benghalensis*	r?
	Black-winged Stilt *Himantopus himantopus*	m 5
	Northern Stone-curlew *Burhinus oedicnemus*	r 3
	Little Pratincole *Glareola lactea*	r?
	Little Ringed Plover *Charadrius dubius*	r w? 3
	River Plover *Hoplopterus duvaucelii*	r 3
	Yellow-wattled Plover *Hoplopterus malabaricus*	r 3
	Red-wattled Plover *Hoplopterus indicus*	r 1
	Temminck's Stint *Calidris temminckii*	w 2
	Long-toed Stint *Calidris subminuta*	v
	Ruff *Philomachus pugnax*	m
	Common Snipe *Gallinago gallinago*	w 3
	Spotted Redshank *Tringa erythropus*	m
	Marsh Sandpiper *Tringa stagnatilis*	m
	Common Greenshank *Tringa nebularia*	w 2
	Green Sandpiper *Tringa ochropus*	w 2
	Wood Sandpiper *Tringa glareola*	m
	Common Sandpiper *Actitis hypoleucos*	w m 1
	River Tern *Sterna aurantia*	r? m?
	Black-bellied Tern *Sterna acuticauda*	r 3
	Rock Pigeon *Columba livia*	r 2
	Eurasian Collared Dove *Streptopelia decaocto*	r 1
	Red Turtle Dove *Streptopelia tranquebarica*	r b 2
	Oriental Turtle Dove *Streptopelia orientalis*	w? 2
	Spotted Dove *Streptopelia chinensis*	r 1
	Emerald Dove *Chalcophaps indica*	r 2
R	Orange-breasted Green Pigeon *Treron bicincta*	r 4
	Yellow-footed Green Pigeon *Treron phoenicoptera*	r 3
	Alexandrine Parakeet *Psittacula eupatria*	r 1

	Ring-necked Parakeet *Psittacula krameri*	r b 1
	Blossom-headed Parakeet *Psittacula cyanocephala*	r 3
	Moustached Parakeet *Psittacula alexandri*	r? 2
	Pied Crested Cuckoo *Clamator jacobinus*	m? s? 5
	Common Hawk-Cuckoo *Hierococcyx varius*	s 2
	Indian Cuckoo *Cuculus micropterus*	s 3
	Common Cuckoo *Cuculus canorus*	s
	Common Koel *Eudynamys scolopacea*	r? s 3
	Sirkeer Malkoha *Phaenicophaeus leschenaultii*	r 3
	Greater Coucal *Centropus sinensis*	r 2
	Lesser Coucal *Centropus bengalensis*	r 3
E	Grass Owl *Tyto capensis*	r? 5
	Indian Scops Owl *Otus bakkamoena*	r 3
	Oriental Scops Owl *Otus sunia*	r 2
V	Brown Fish Owl *Ketupa zeylonensis*	r 3
	Jungle Owlet *Glaucidium radiatum*	r 1
	Asian Barred Owlet *Glaucidium cuculoides*	r
	Brown Hawk Owl *Ninox scutulata*	r 3
	Spotted Little Owl *Athene brama*	r b 1
	Savanna Nightjar *Caprimulgus affinis*	r ? 3
	Indian Nightjar *Caprimulgus asiaticus*	r ? 5
	Large-tailed Nightjar *Caprimulgus macrurus*	r ? 2
R	White-rumped Needletail *Zoonavena sylvatica*	r 4
	Alpine Swift *Apus melba*	m 3
	Little Swift *Apus affinis*	m 5
	Crested Tree Swift *Hemiprocne coronata*	r 4
	White-breasted Kingfisher *Halcyon smyrnensis*	r 2
	Stork-billed Kingfisher *Pelargopsis capensis*	r 3
	Common Kingfisher *Alcedo atthis*	r 2
	Pied Kingfisher *Ceryle rudis*	r 3
	Green Bee-eater *Merops orientalis*	r? s 2
	Blue-tailed Bee-eater *Merops philippinus*	s 3
	Indian Roller *Coracias benghalensis*	r 1
R	Dollarbird *Eurystomus orientalis*	s 5
	Hoopoe *Upupa epops*	r 3
	Indian Grey Hornbill *Tockus birostris*	r 2
V	Oriental Pied Hornbill *Anthracoceros coronatus*	r 3
	Brown-headed Barbet *Megalaima zeylanica*	r 3
	Coppersmith Barbet *Megalaima haemacephala*	r 3
	Rufous Woodpecker *Celeus brachyurus*	r 3
	Grey-headed Woodpecker *Picus canus*	r 2
+	Streak-throated Green Woodpecker *Picus myrmecophoneus*	r 3
	Himalayan Golden-backed Woodpecker *Dinopium shorii*	r 3
	Lesser Golden-backed Woodpecker *Dinopium benghalense*	r 2
	Greater Golden-backed Woodpecker *Chrysocolaptes lucidus*	r
	White-naped Woodpecker *Chrysocolaptes festivus*	r 3
	Bay Woodpecker *Blythipicus pyrrhotis*	v?
V	Great Slaty Woodpecker *Mulleripicus pulverulentus*	r 3
	Yellow-crowned Pied Woodpecker *Dendrocopos mahrattensis*	r 3
	Brown-capped Pygmy Woodpecker *Dendrocopos moluccensis*	r 2
R	Indian Pitta *Pitta brachyura*	s 5
	Bengal Bush Lark *Mirafra assamica*	r 2

	Ashy-crowned Finchlark *Eremopterix grisea*	r 2
	Oriental Skylark *Alauda gulgula*	r ? 3
	Brown-throated Sand Martin *Riparia paludicola*	r 2
	Barn Swallow *Hirundo rustica*	w m 3
	Red-rumped Swallow *Hirundo daurica*	w m 2
+	Nepal House-Martin *Delichon nipalensis*	w 3
	Richard's Pipit *Anthus novaeseelandiae*	r 2
	Olive-backed Pipit *Anthus hodgsoni*	w 2
	Yellow Wagtail *Motacilla flava*	w 3
	Citrine Wagtail *Motacilla citreola*	w 3
	Grey Wagtail *Motacilla cinerea*	w 3
	White Wagtail *Motacilla alba*	w 2
	White-browed Wagtail *Motacilla maderaspatensis*	r w? 3
	Bar-winged Flycatcher-shrike *Hemipus picatus*	r 2
	Black-winged Cuckoo-shrike *Coracina melaschistos*	r
	Large Cuckoo-shrike *Coracina novaehollandiae*	r 3
	Scarlet Minivet *Pericrocotus flammeus*	w 2
	Long-tailed Minivet *Pericrocotus ethologus*	w 3
	Small Minivet *Pericrocotus cinnamomeus*	r? 2
	Black-crested Bulbul *Pycnonotus melanicterus*	r 3
	Red-whiskered Bulbul *Pycnonotus jocosus*	r 3
	Red-vented Bulbul *Pycnonotus cafer*	r 2
	Common Iora *Aegithina tiphia*	r w 3
	Golden-fronted Leafbird *Chloropsis aurifrons*	r 3
	Siberian Rubythroat *Luscinia calliope*	w 5
	Bluethroat *Luscinia svecica*	w m 2
	Asian Magpie-Robin *Copsychus saularis*	r 3
	White-rumped Shama *Copsychus malabaricus*	r 3
	Common Stonechat *Saxicola torquata*	r 1
R+	White-tailed Stonechat *Saxicola leucura*	r b 3
	Pied Bushchat *Saxicola caprata*	r b 1
	Grey Bushchat *Saxicola ferrea*	w 5
	Indian Robin *Saxicoloides fulicata*	r 3
	Blue Whistling Thrush *Myiophoneus caeruleus*	w 3
	Orange-headed Ground Thrush *Zoothera citrina*	w 3
+	Tickell's Thrush *Turdus unicolor*	w 3
	Dark-throated Thrush *Turdus ruficollis*	w 3
R*	Pale-footed Bush Warbler *Cettia pallidipes*	r 3
	Spotted Bush Warbler *Bradypterus thoracicus*	w 5
	Brown Bush Warbler *Bradypterus luteoventris*	v
R	Bright-capped Cisticola *Cisticola exilis*	r? 2
	Fantail Cisticola *Cisticola juncidis*	r 2
	Plain Prinia *Prinia inornata*	r 3
	Ashy Prinia *Prinia socialis*	r 2
	Grey-breasted Prinia *Prinia hodgsoni*	r? 2
	Yellow-bellied Prinia *Prinia flaviventris*	r 2
	Jungle Prinia *Prinia sylvatica*	r 3
V	Large Grass Warbler *Graminicola bengalensis*	r 5
	Common Tailorbird *Orthotomus sutorius*	r 2
	Lanceolated Warbler *Locustella lanceolata*	v
V	Striated Marsh Warbler *Megalurus palustris*	r
	Blyth's Reed Warbler *Acrocephalus dumetorum*	w m 2

	Orphean Warbler *Sylvia hortensis*	v
	Golden-spectacled Warbler *Seicercus burkii*	w 3
+	Grey-hooded Warbler *Seicercus xanthoschistos*	w 2
	Blyth's Crowned Warbler *Phylloscopus reguloides*	w
	Greenish Warbler *Phylloscopus trochiloides*	w m
	Pallas's Leaf Warbler *Phylloscopus proregulus*	w 3
	Yellow-browed Warbler *Phylloscopus inornatus*	w 3
	Dusky Warbler *Phylloscopus fuscatus*	w 5
*	Smoky Warbler *Phylloscopus fuligiventer*	w 5
	Tickell's Warbler *Phylloscopus affinis*	w 3
	Chiffchaff *Phylloscopus collybita*	w 3
	Tickell's Blue Flycatcher *Cyornis tickelliae*	r? w 2
+	Rufous-tailed Flycatcher *Muscicapa ruficauda*	w 3
	Slaty-blue Flycatcher *Ficedula tricolor*	w 3
R	Little Pied Flycatcher *Ficedula westermanni*	w 5
	Grey-headed Flycatcher *Culicicapa ceylonensis*	w 2
	White-throated Fantail *Rhipidura albicollis*	r 2
	White-browed Fantail *Rhipidura aureola*	r 3
	Asian Paradise Flycatcher *Terpsiphone paradisi*	s 2
	Black-naped Monarch *Hypothymis azurea*	r 3
	Striped Tit-Babbler *Macronous gularis*	r 3
	Red-capped Babbler *Timalia pileata*	r 3
	Yellow-eyed Babbler *Chrysomma sinense*	r
	Striated Babbler *Turdoides earlei*	r 2
	Jungle Babbler *Turdoides striatus*	r 2
	Great Tit *Parus major*	r 1
	Chestnut-bellied Nuthatch *Sitta castanea*	r 2
	Purple Sunbird *Nectarinia asiatica*	r 1
	Crimson Sunbird *Aethopyga siparaja*	r 3
	Thick-billed Flowerpecker *Dicaeum agile*	r? w b 3
	Pale-billed Flowerpecker *Dicaeum erythrorhynchos*	r 2
	Oriental White-eye *Zosterops palpebrosa*	r 2
	Black-hooded Oriole *Oriolus xanthornus*	r 1
	Golden Oriole *Oriolus oriolus*	s
	Long-tailed Shrike *Lanius schach*	w 2
+	Grey-backed Shrike *Lanius tephronotus*	w 3
	Black Drongo *Dicrurus macrocercus*	r 1
	White-bellied Drongo *Dicrurus caerulescens*	r 1
	Spangled Drongo *Dicrurus hottentottus*	r 3
	Greater Racket-tailed Drongo *Dicrurus paradiseus*	r 1
	Rufous Treepie *Dendrocitta vagabunda*	r 2
	House Crow *Corvus splendens*	r 3
	Jungle Crow *Corvus macrorhynchos*	r 1
	Chestnut-tailed Starling *Sturnus malabaricus*	w 3
	Brahminy Starling *Sturnus pagodarum*	r 3
	Common Starling *Sturnus vulgaris*	m 4
	Asian Pied Starling *Sturnus contra*	r 3
	Common Mynah *Acridotheres tristis*	r 2
	Bank Mynah *Acridotheres ginginianus*	r 2
	Jungle Mynah *Acridotheres fuscus*	r 3
	House Sparrow *Passer domesticus*	r b 2
	Yellow-throated Sparrow *Petronia xanthocollis*	r 3

R+	Black-breasted Weaver *Ploceus benghalensis*	r 2
	Streaked Weaver *Ploceus manyar*	r 4
	Baya Weaver *Ploceus philippinus*	r w 1
	Red Avadavat *Amandava amandava*	r 2
	Yellow-breasted Greenfinch *Carduelis spinoides*	w
	Scaly-breasted Munia *Lonchura punctulata*	r 2
	Common Rosefinch *Carpodacus erythrinus*	w 3
	Yellow-breasted Bunting *Emberiza aureola*	w 3
	Crested Bunting *Melophus lathami*	r? w 3

ROYAL BARDIA WILDLIFE RESERVE

	Great Cormorant *Phalacrocorax carbo*	r
	Little Cormorant *Phalacrocorax niger*	w? m?
	Darter *Anhinga melanogaster*	r?
	Cinnamon Bittern *Ixobrychus cinnamomeus*	s?
	Black-crowned Night Heron *Nycticorax nycticorax*	r? s?
	Green-backed Heron *Butoridus striatus*	r? s?
	Indian Pond Heron *Ardeola grayii*	r
	Cattle Egret *Bubulcus ibis*	r
	Little Egret *Egretta garzetta*	r
	Intermediate Egret *Egretta intermedia*	r
	Great Egret *Egretta alba*	r
	Grey Heron *Ardea cinerea*	w
	Purple Heron *Ardea purpurea*	r
	Black Stork *Ciconia nigra*	w? m?
	Woolly-necked Stork *Ciconia episcopus*	r? s?
	Black-necked Stork *Xenorhynchus asiaticus*	r? w?
	Lesser Adjutant Stork *Leptoptilos javanicus*	r
	Red-naped Ibis *Pseudibis papillosa*	r
	Lesser Whistling Duck *Dendrocygna javanica*	w m?
	Greylag Goose *Anser anser*	w? m?
	Bar-headed Goose *Anser indicus*	m
	Ruddy Shelduck *Tadorna ferruginea*	w
	Mallard *Anas platyrhynchos*	w? m?
	Northern Pintail *Anas acuta*	w? m?
	Goosander *Mergus merganser*	w
	Crested Honey Buzzard *Pernis ptilorhyncus*	r? m?
	Black-shouldered Kite *Elanus caeruleus*	r
	Black Kite *Milvus migrans*	r
	Brahminy Kite *Haliastur indus*	r
E	Pallas's Fish Eagle *Haliaeetus leucoryphus*	w? m?
	White-tailed Eagle *Haliaeetus albicilla*	w
E	Lesser Fishing Eagle *Ichthyophaga nana*	r
	Egyptian Vulture *Neophron percnopterus*	r
	Lammergeier *Gypaetus barbatus*	m?
	Oriental White-backed Vulture *Gyps bengalensis*	r
	Long-billed Vulture *Gyps indicus*	r
	Eurasian Griffon Vulture *Gyps fulvus*	r
	Red-headed Vulture *Sarcogyps calvus*	r
	Crested Serpent Eagle *Spilornis cheela*	r? w?
	Marsh Harrier *Circus aeruginosus*	w? m?
	Hen Harrier *Circus cyaneus*	w? m?
	Pied Harrier *Circus melanoleucus*	w? m?
	Northern Sparrowhawk *Accipiter nisus*	r? w?
	Shikra *Accipiter badius*	r
	White-eyed Buzzard *Butastur teesa*	r
R	Tawny Eagle *Aquila rapax vindhiana*	r
	Steppe Eagle *Aquila rapax nipalensis*	w? m
E	Rufous-bellied Eagle *Hieraaetus kienerii*	r? m?
V	Changeable Hawk-Eagle *Spizaetus cirrhatus*	r
	Mountain Hawk-Eagle *Spizaetus nipalensis*	w?

	Osprey *Pandion haliaetus*	r? w?
	Common Kestrel *Falco tinnunculus*	r? w? m?
	Eurasian Hobby *Falco subbuteo*	r? m? w?
	Black Francolin *Francolinus francolinus*	r
	Grey Francolin *Francolinus pondicerianus*	r
	Red Junglefowl *Gallus gallus*	r
	Blue Peafowl *Pavo cristatus*	r
	Barred Buttonquail *Turnix suscitator*	r
	White-breasted Waterhen *Amaurornis phoenicurus*	r
	Common Moorhen *Gallinula chloropus*	r? w?
	Sarus Crane *Grus antigone*	r
E*	Bengal Florican *Houbaropsis bengalensis*	r?
E	Lesser Florican *Sypheotides indica*	s?
	Northern Stone-curlew *Burhinus oedicnemus*	r
	Great Stone-Plover *Esacus recurvirostris*	r? w?
	Little Pratincole *Glareola lactea*	r?
	Little Ringed Plover *Charadrius dubius*	r? w?
	Kentish Plover *Charadrius alexandrinus*	w? m?
	River Plover *Hoplopterus duvaucelii*	r b
	Red-wattled Plover *Hoplopterus indicus*	r
	Temminck's Stint *Calidris temminckii*	w m
	Common Redshank *Tringa totanus*	w m
	Common Greenshank *Tringa nebularia*	w m
	Common Sandpiper *Actitis hypoleucos*	w m
	Great Black-headed Gull *Larus ichthyaetus*	w m
	Brown-headed Gull *Larus brunnicephalus*	w m
	Caspian Tern *Sterna caspia*	m
	River Tern *Sterna aurantia*	r
	Black-bellied Tern *Sterna acuticauda*	r? s?
	Little Tern *Sterna albifrons*	s b
	Collared Dove *Streptopelia tranquebarica*	r
	Red Turtle Dove *Streptopelia decaocto*	r
	Oriental Turtle Dove *Streptopelia orientalis*	w
	Spotted Dove *Streptopelia chinensis*	r
	Emerald Dove *Chalcophaps indica*	r
R	Orange-breasted Green Pigeon *Treron bicincta*	r
	Yellow-footed Green Pigeon *Treron phoenicoptera*	r
V	Pin-tailed Green Pigeon *Treron apicauda*	r
	Alexandrine Parakeet *Psittacula eupatria*	r
	Ring-necked Parakeet *Psittacula krameri*	r
	Blossom-headed Parakeet *Psittacula cyanocephala*	w?
	Slaty-headed Parakeet *Psittacula himalayana*	w?
	Grey-bellied Plaintive Cuckoo *Cacomantis passerinus*	s
	Common Hawk-Cuckoo *Hierococcyx varius*	s
	Indian Cuckoo *Cuculus micropterus*	s
	Common Cuckoo *Cuculus canorus*	s
	Drongo-Cuckoo *Surniculus lugubris*	s
	Common Koel *Eudynamys scolopacea*	r
	Green-billed Malkoha *Phaenicophaeus tristis*	r
	Sirkeer Malkoha *Phaenicophaeus leschenaultii*	r
	Greater Coucal *Centropus sinensis*	r
	Lesser Coucal *Centropus bengalensis*	r? s?

E	Forest Eagle Owl *Bubo nipalensis*	r
V	Brown Fish Owl *Ketupa zeylonensis*	r
	Jungle Owlet *Glaucidium radiatum*	r
	Asian Barred Owlet *Glaucidium cuculoides*	r
	Brown Hawk Owl *Ninox scutulata*	r
	Spotted Little Owl *Athene brama*	r
	Savanna Nightjar *Caprimulgus affinis*	?
	Indian Nightjar *Caprimulgus asiaticus*	?
	Large-tailed Nightjar *Caprimulgus macrurus*	?
	Jungle Nightjar *Caprimulgus indicus*	r
	Himalayan Swiftlet *Collocalia brevirostris*	r
R	White-rumped Needletail *Zoonavena sylvatica*	r?
	Alpine Swift *Apus melba*	?
	Little Swift *Apus affinis*	r
	Crested Tree Swift *Hemiprocne coronata*	r
	White-breasted Kingfisher *Halcyon smyrnensis*	r
	Stork-billed Kingfisher *Pelargopsis capensis*	r
	Common Kingfisher *Alcedo atthis*	r
	Pied Kingfisher *Ceryle rudis*	r
	Blue-bearded Bee-eater *Nyctyornis athertoni*	r
	Green Bee-eater *Merops orientalis*	r s
	Blue-tailed Bee-eater *Merops philippinus*	s
	Chestnut-headed Bee-eater *Merops leschenaulti*	s
	Indian Roller *Coracias benghalensis*	r
R	Dollarbird *Eurystomus orientalis*	s
	Hoopoe *Upupa epops*	r m
	Indian Grey Hornbill *Tockus birostris*	r
V	Oriental Pied Hornbill *Anthracoceros coronatus*	r
E	Great Pied Hornbill *Buceros bicornis*	r
	Brown-headed Barbet *Megalaima zeylanica*	r
	Lineated Barbet *Megalaima lineata*	r
	Blue-throated Barbet *Megalaima asiatica*	r
	Coppersmith Barbet *Megalaima haemacephala*	r
	Rufous Woodpecker *Celeus brachyurus*	r
	Grey-headed Woodpecker *Picus canus*	r
	Streak-throated Green Woodpecker *Picus myrmecophoneus*	r
	Himalayan Golden-backed Woodpecker *Dinopium shorii*	r
	Lesser Golden-backed Woodpecker *Dinopium benghalense*	r
	Greater Golden-backed Woodpecker *Chrysocolaptes lucidus*	r
	White-naped Woodpecker *Chrysocolaptes festivus*	r
V	Great Slaty Woodpecker *Mulleripicus pulverulentus*	r
	Yellow-crowned Pied Woodpecker *Dendrocopos mahrattensis*	r
	Brown-capped Pygmy Woodpecker *Dendrocopos moluccensis*	r
R	Indian Pitta *Pitta brachyura*	s
	Bengal Bush Lark *Mirafra assamica*	r
	Ashy-crowned Finchlark *Eremopterix grisea*	r
	Sandlark *Calandrella raytal*	r
	Crested Lark *Galerida cristata*	r
	Brown-throated Sand Martin *Riparia paludicola*	r
	Barn Swallow *Hirundo rustica*	r? w?
+	Nepal House-Martin *Delichon nipalensis*	r
	Olive-backed Pipit *Anthus hodgsoni*	w

	Richard's Pipit *Anthus novaeseelandiae*	r? w? m?
	Yellow Wagtail *Motacilla flava*	w? m?
	Grey Wagtail *Motacilla cinerea*	w
	White Wagtail *Motacilla alba*	w? m?
	White-browed Wagtail *Motacilla maderaspatensis*	r
	Common Woodshrike *Tephrodornis pondicerianus*	r
	Bar-winged Flycatcher-shrike *Hemipus picatus*	r
	Black-headed Cuckoo-shrike *Coracina melanoptera*	s?
	Black-winged Cuckoo-shrike *Coracina melaschistos*	r
	Large Cuckoo-shrike *Coracina novaehollandiae*	r
	Scarlet Minivet *Pericrocotus flammeus*	r
	Small Minivet *Pericrocotus cinnamomeus*	r
R	Rosy Minivet *Pericrocotus roseus*	r?
+	Striated Bulbul *Pycnonotus striatus*	r
	Black-crested Bulbul *Pycnonotus melanicterus*	r
	Red-whiskered Bulbul *Pycnonotus jocosus*	r b
	White-cheeked Bulbul *Pycnonotus leucogenys*	r
	Red-vented Bulbul *Pycnonotus cafer*	r
	Black Bulbul *Hypsipetes madagascariensis*	r
	Common Iora *Aegithina tiphia*	r
	Golden-fronted Leafbird *Chloropsis aurifrons*	r
	White-tailed Rubythroat *Luscinia pectoralis*	w? m?
	Asian Magpie-Robin *Copsychus saularis*	r
	White-rumped Shama *Copsychus malabaricus*	r
	Hodgson's Redstart *Phoenicurus hodgsoni*	w
	White-bellied Redstart *Hodgsonius phoenicuroides*	m
	Common Stonechat *Saxicola torquata*	r? w? m?
R+	White-tailed Stonechat *Saxicola leucura*	r
	Pied Bushchat *Saxicola caprata*	r
	Grey Bushchat *Saxicola ferrea*	r?
	White-capped Redstart *Chaimarrornis leucocephalus*	w
	Indian Robin *Saxicoloides fulicata*	r
	Blue Whistling Thrush *Myiophoneus caeruleus*	w?
	Dark-throated Thrush *Turdus ruficollis*	w
	Grey-bellied Tesia *Tesia cyaniventer*	w
R*	Pale-footed Bush Warbler *Cettia pallidipes*	r
	Fantail Cisticola *Cisticola juncidis*	r
	Graceful Prinia *Prinia gracilis*	r
	Plain Prinia *Prinia inornata*	r
	Ashy Prinia *Prinia socialis*	s r
	Grey-breasted Prinia *Prinia hodgsoni*	r
	Jungle Prinia *Prinia sylvatica*	r
*	Grey-capped Prinia *Prinia cinereocapilla*	r
	Common Tailorbird *Orthotomus sutorius*	r
	Blyth's Reed Warbler *Acrocephalus dumetorum*	w
	Golden-spectacled Warbler *Seicercus burkii*	w
	Chestnut-crowned Warbler *Seicercus castaniceps*	w?
+	Grey-hooded Warbler *Seicercus xanthoschistos*	r?
	Greenish Warbler *Phylloscopus trochiloides*	w? m?
	Yellow-browed Warbler *Phylloscopus inornatus*	w? m?
	Tickell's Blue Flycatcher *Cyornis tickelliae*	w? m?
	Verditer Flycatcher *Muscicapa thalassina*	r?

	Asian Sooty Flycatcher *Muscicapa sibirica*	m
	Asian Brown Flycatcher *Muscicapa latirostris*	m
	Slaty-blue Flycatcher *Ficedula tricolor*	w? m?
	Snowy-browed Flycatcher *Ficedula hyperythra*	w?
	Orange-gorgetted Flycatcher *Ficedula strophiata*	w
	Grey-headed Flycatcher *Culicicapa ceylonensis*	w?
+	Yellow-bellied Fantail *Rhipidura hypoxantha*	w?
	White-browed Fantail *Rhipidura aureola*	r
	Asian Paradise Flycatcher *Terpsiphone paradisi*	s
	Black-naped Monarch *Hypothymis azurea*	r
	Striped Tit-Babbler *Macronous gularis*	r
	Yellow-eyed Babbler *Chrysomma sinense*	r
	Spiny Babbler *Turdoides nipalensis*	r
	Striated Babbler *Turdoides earlei*	r
	Jungle Babbler *Turdoides striatus*	r
+	White-throated Laughing-thrush *Garrulax albogularis*	r
E	Silver-eared Mesia *Leiothrix argentauris*	r
	Great Tit *Parus major*	r
	Velvet-fronted Nuthatch *Sitta frontalis*	r
	Chestnut-bellied Nuthatch *Sitta castanea*	r
	Wallcreeper *Tichodroma muraria*	w
	Bar-tailed Treecreeper *Certhia himalayana*	w
	Purple Sunbird *Nectarinia asiatica*	r
	Crimson Sunbird *Aethopyga siparaja*	r
	Thick-billed Flowerpecker *Dicaeum agile*	r
	Pale-billed Flowerpecker *Dicaeum erythrorhynchos*	r
	Buff-bellied Flowerpecker *Dicaeum ignipectus*	r
	Oriental White-eye *Zosterops palpebrosa*	r
	Black-hooded Oriole *Oriolus xanthornus*	r
	Golden Oriole *Oriolus oriolus*	s
	Brown Shrike *Lanius cristatus*	w? m?
	Bay-backed Shrike *Lanius vittatus*	w? m?
	Long-tailed Shrike *Lanius schach*	r
	Great Grey Shrike *Lanius excubitor*	r
	Black Drongo *Dicrurus macrocercus*	r
	Ashy Drongo *Dicrurus leucophaeus*	r
	White-bellied Drongo *Dicrurus caerulescens*	r
V	Crow-billed Drongo *Dicrurus annectans*	s?
	Bronzed Drongo *Dicrurus aeneus*	r
	Spangled Drongo *Dicrurus hottentottus*	r
	Greater Racket-tailed Drongo *Dicrurus paradiseus*	r
+	Lanceolated Jay *Garrulus lanceolatus*	r?
	Red-billed Blue Magpie *Urocissa erythrorhyncha*	r
	Rufous Treepie *Dendrocitta vagabunda*	r
	House Crow *Corvus splendens*	r
	Jungle Crow *Corvus macrorhynchos*	r
	Chestnut-tailed Starling *Sturnus malabaricus*	r
	Asian Pied Starling *Sturnus contra*	r
	Common Mynah *Acridotheres tristis*	r
	Jungle Mynah *Acridotheres fuscus*	r
	House Sparrow *Passer domesticus*	r
	Eurasian Tree Sparrow *Passer montanus*	r

Baya Weaver *Ploceus philippinus*   r
Red Avadavat *Amandava amandava*   r
Striated Munia *Lonchura striata*   r
Scaly-breasted Munia *Lonchura punctulata*   r
Common Rosefinch *Carpodacus erythrinus*   w
Crested Bunting *Melophus lathami*   r

KOSI TAPPU WILDLIFE RESERVE

Little Grebe *Tachybaptus ruficollis*	r w m
Great Cormorant *Phalacrocorax carbo*	r
Little Cormorant *Phalacrocorax niger*	w m
Darter *Anhinga melanogaster*	r
Spot-billed Pelican *Pelecanus philippensis*	w m
Eurasian Bittern *Botaurus stellaris*	w m
Cinnamon Bittern *Ixobrychus cinnamomeus*	s
Black-crowned Night Heron *Nycticorax nycticorax*	r s
Green-backed Heron *Butorides striatus*	r s
Indian Pond Heron *Ardeola grayii*	r
Cattle Egret *Bubulcus ibis*	r
Little Egret *Egretta garzetta*	r
Intermediate Egret *Egretta intermedia*	r
Great Egret *Egretta alba*	r
Grey Heron *Ardea cinerea*	w
Purple Heron *Ardea purpurea*	r
Painted Stork *Mycteria leucocephala*	s?
Asian Openbill Stork *Anastomus oscitans*	r m
Black Stork *Ciconia nigra*	w m
Woolly-necked Stork *Ciconia episcopus*	r
Black-necked Stork *Xenorhynchus asiaticus*	r w
Greater Adjutant Stork *Leptoptilos dubius*	s?
Lesser Adjutant Stork *Leptoptilos javanicus*	r
Red-naped Ibis *Pseudibis papillosa*	r b
Oriental White Ibis *Threskiornis melanocephalus*	r
Lesser Whistling Duck *Dendrocygna javanica*	w m
Greylag Goose *Anser anser*	w m
Bar-headed Goose *Anser indicus*	w m
Ruddy Shelduck *Tadorna ferruginea*	w
Cotton Pygmy Goose *Nettapus coromandelianus*	r s
Eurasian Wigeon *Anas penelope*	w m
Falcated Duck *Anas falcata*	w m
Gadwall *Anas strepera*	w m
Common Teal *Anas crecca*	w m
Mallard *Anas platyrhynchos*	w m
Spotbill *Anas poecilorhyncha*	r w
Northern Pintail *Anas acuta*	w m
Garganey *Anas querquedula*	w m
Northern Shoveler *Anas clypeata*	w m
Common Pochard *Aythya ferina*	w m
Ferruginous Duck *Aythya nyroca*	w m
Tufted Duck *Aythya fuligula*	w m
Goosander *Mergus merganser*	w
Crested Honey Buzzard *Pernis ptilorhyncus*	r
Black-shouldered Kite *Elanus caeruleus*	r
Black Kite *Milvus migrans*	r m
Brahminy Kite *Haliastur indus*	r
Pallas's Fish Eagle *Haliaeetus leucoryphus*	w m
White-tailed Eagle *Haliaeetus albicilla*	w m
Egyptian Vulture *Neophron percnopterus*	r

	Oriental White-backed Vulture *Gyps bengalensis*	r
	Long-billed Vulture *Gyps indicus*	r
	Eurasian Griffon Vulture *Gyps fulvus*	r
	Red-headed Vulture *Sarcogyps calvus*	w m
	Eurasian Black Vulture *Aegypius monachus*	w
	Short-toed Eagle *Circaetus gallicus*	w
	Crested Serpent Eagle *Spilornis cheela*	w
	Eurasian Marsh Harrier *Circus aeruginosus*	w m
	Hen Harrier *Circus cyaneus*	w m
	Montagu's Harrier *Circus pygargus*	w m
	Pied Harrier *Circus melanoleucus*	w m
	Shikra *Accipiter badius*	r
	White-eyed Buzzard *Butastur teesa*	r
	Common Buzzard *Buteo buteo*	w m
	Long-legged Buzzard *Buteo rufinus*	w m
	Black Eagle *Ictinaetus malayensis*	w
R	Lesser Spotted Eagle *Aquila pomarina*	r
	Steppe Eagle *Aquila rapax nipalensis*	w m
R	Tawny Eagle *Aquila rapax vindhiana*	r
	Booted Eagle *Hieraaetus pennatus*	w m
E	Rufous-bellied Eagle *Hieraaetus kienerii*	m?
V	Changeable Hawk-Eagle *Spizaetus cirrhatus*	r
	Osprey *Pandion haliaetus*	r w
	Red-thighed Falconet *Microhierax caerulescens*	r
	Common Kestrel *Falco tinnunculus*	r w m
E	Red-necked Falcon *Falco chicquera*	r?
	Eurasian Hobby *Falco subbuteo*	m
	Peregrine *Falco peregrinus*	w m
	Black Francolin *Francolinus francolinus*	r
V*	Swamp Francolin *Francolinus gularis*	r
	Red Junglefowl *Gallus gallus*	r
	Blue Peafowl *Pavo cristatus*	r
R	Yellow-legged Buttonquail *Turnix tanki*	r
	Barred Buttonquail *Turnix suscitator*	r
	Brown Crake *Amaurornis akool*	r?
	White-breasted Waterhen *Amaurornis phoenicurus*	r
	Common Moorhen *Gallinula chloropus*	r w
V	Watercock *Gallicrex cinerea*	s
	Purple Gallinule *Porphyrio porphyrio*	w m
E*	Bengal Florican *Houbaropsis bengalensis*	r?
	Pheasant-tailed Jacana *Hydrophasianus chirurgus*	r? s
	Bronze-winged Jacana *Metopidius indicus*	r
	Black-winged Stilt *Himantopus himantopus*	m
	Northern Stone-curlew *Burhinus oedicnemus*	r
	Great Stone-Plover *Esacus recurvirostris*	r w
	Little Pratincole *Glareola lactea*	r
	Little Ringed Plover *Charadrius dubius*	r w
	Kentish Plover *Charadrius alexandrinus*	w m
	Pacific Golden Plover *Pluvialis fulva*	w m
	River Plover *Hoplopterus duvaucelii*	r
	Yellow-wattled Plover *Hoplopterus malabaricus*	r
	Red-wattled Plover *Hoplopterus indicus*	r

	Little Stint *Calidris minuta*	w?	m?
	Temminck's Stint *Calidris temminckii*	w	m
	Common Snipe *Gallinago gallinago*	w?	m?
	Pintail Snipe *Gallinago stenura*	w	m
	Spotted Redshank *Tringa erythropus*	w?	m?
	Common Redshank *Tringa totanus*	w	m
	Common Greenshank *Tringa nebularia*	w	m
	Green Sandpiper *Tringa ochropus*	w	m
	Wood Sandpiper *Tringa glareola*	w	m
	Common Sandpiper *Actitis hypoleucos*	w	m
	Brown-headed Gull *Larus brunnicephalus*	w	m
	River Tern *Sterna aurantia*	r	m
	Common Tern *Sterna hirundo*	m	
	Black-bellied Tern *Sterna acuticauda*	r	s m
	Little Tern *Sterna albifrons*	s	
	Whiskered Tern *Chlidonias hybridus*	w	m
	Rock Pigeon *Columba livia*	r	
	Collared Dove *Streptopelia decaocto*	r	
	Red Turtle Dove *Streptopelia tranquebarica*	r	
	Oriental Turtle Dove *Streptopelia orientalis*	w	
	Laughing Dove *Streptopelia senegalensis*	?	
	Spotted Dove *Streptopelia chinensis*	r	
	Emerald Dove *Chalcophaps indica*	r	
R	Orange-breasted Green Pigeon *Treron bicincta*	r	
R	Pompadour Green Pigeon *Treron pompadora*	r	
	Yellow-footed Green Pigeon *Treron phoenicoptera*	r	
	Ring-necked Parakeet *Psittacula krameri*	r	
+	Slaty-headed Parakeet *Psittacula himalayana*	w	
	Blossom-headed Parakeet *Psittacula cyanocephala*	r	
	Pied Crested Cuckoo *Clamator jacobinus*	s	
	Hodgson's Hawk-Cuckoo *Hierococcyx fuggax*	v	
	Common Hawk-Cuckoo *Hierococcyx varius*	s	
	Grey-bellied Plaintive Cuckoo *Cacomantis passerinus*	s	
	Indian Cuckoo *Cuculus micropterus*	s	
	Common Cuckoo *Cuculus canorus*	s	
	Oriental Cuckoo *Cuculus saturatus*	m	
	Drongo-Cuckoo *Surniculus lugubris*	s	
	Common Koel *Eudynamys scolopacea*	r?	
	Greater Coucal *Centropus sinensis*	r	
	Lesser Coucal *Centropus bengalensis*	r	
E	Dusky Eagle Owl *Bubo coromandus*	r	
V	Brown Fish Owl *Ketupa zeylonensis*	r?	
	Jungle Owlet *Glaucidium radiatum*	r	
	Brown Hawk Owl *Ninox scutulata*	r	
	Spotted Little Owl *Athene brama*	r	
	Short-eared Owl *Asio flammeus*	w?	m?
	Indian Nightjar *Caprimulgus asiaticus*	?	
	Large-tailed Nightjar *Caprimulgus macrurus*	?	
	Little Swift *Apus affinis*	r	
	Crested Tree Swift *Hemiprocne coronata*	r?	
	White-breasted Kingfisher *Halcyon smyrnensis*	r	
	Stork-billed Kingfisher *Pelargopsis capensis*	r	

	Common Kingfisher *Alcedo atthis*	r
	Pied Kingfisher *Ceryle rudis*	r
	Green Bee-eater *Merops orientalis*	r s
	Blue-tailed Bee-eater *Merops philippinus*	s
	Chestnut-headed Bee-eater *Merops leschenaulti*	r? s
	Indian Roller *Coracias benghalensis*	r
	Hoopoe *Upupa epops*	r m
	Indian Grey Hornbill *Tockus birostris*	r
	Blue-throated Barbet *Megalaima asiatica*	r
	Coppersmith Barbet *Megalaima haemacephala*	r
	Eurasian Wryneck *Jynx torquilla*	w m
	Rufous Woodpecker *Celeus brachyurus*	r
	Grey-headed Woodpecker *Picus canus*	r
	Streak-throated Green Woodpecker *Picus myrmecophoneus*	r
	Lesser Golden-backed Woodpecker *Dinopium benghalense*	r
	Fulvous-breasted Pied Woodpecker *Dendrocopos macei*	r
	Bengal Bushlark *Mirafra assamica*	r
	Ashy-crowned Finchlark *Eremopterix grisea*	r
	Sandlark *Calandrella raytal*	r
	Greater Short-toed Lark *Calandrella brachydactyla*	m
	Crested Lark *Galerida cristata*	r
	Oriental Skylark *Alauda gulgula*	r? w
	Brown-throated Sand Martin *Riparia paludicola*	r
	Collared Sand Martin *Riparia riparia*	m
	Barn Swallow *Hirundo rustica*	r
	Red-rumped Swallow *Hirundo daurica*	w
	Richard's Pipit *Anthus novaeseelandiae*	r w m
	Olive-backed Pipit *Anthus hodgsoni*	m
	Tree Pipit *Anthus trivialis*	m
	Rosy Pipit *Anthus roseatus*	w? m
	Yellow Wagtail *Motacilla flava*	w m
	Citrine Wagtail *Motacilla citreola*	w m
	Grey Wagtail *Motacilla cinerea*	w
	White Wagtail *Motacilla alba*	w m
	White-browed Wagtail *Motacilla maderaspatensis*	r
	Black-headed Cuckoo-shrike *Coracina melanoptera*	s?
	Black-winged Cuckoo-shrike *Coracina melaschistos*	r
	Large Cuckoo-shrike *Coracina novaehollandiae*	r
	Small Minivet *Pericrocotus cinnamomeus*	r
R	Rosy Minivet *Pericrocotus roseus*	s?
	Red-whiskered Bulbul *Pycnonotus jocosus*	r
	Red-vented Bulbul *Pycnonotus cafer*	r
	Common Iora *Aegithina tiphia*	r
	Golden-fronted Leafbird *Chloropsis aurifrons*	r
	Siberian Rubythroat *Luscinia calliope*	w m
	Bluethroat *Luscinia svecica*	w m
	White-tailed Rubythroat *Luscinia pectoralis*	m
	Asian Magpie-Robin *Copsychus saularis*	r
	White-rumped Shama *Copsychus malabaricus*	r
	Black Redstart *Phoenicurus ochruros*	w
	Common Stonechat *Saxicola torquata*	w m
R+	White-tailed Stonechat *Saxicola leucura*	r

	Hodgson's Bushchat *Saxicola insignis*	w
	Pied Bushchat *Saxicola caprata*	r
	Grey Bushchat *Saxicola ferrea*	m?
	White-capped Redstart *Chaimarrornis leucocephalus*	m?
	Blue-capped Rock-Thrush *Monticola cinclorhyncha*	m
	Blue Rock-Thrush *Monticola solitarius*	w? m?
	Blue Whistling Thrush *Myiophoneus caeruleus*	m
	Scaly Thrush *Zoothera dauma*	w
	Orange-headed Ground Thrush *Zoothera citrina*	w? m?
	Tickell's Thrush *Turdus unicolor*	m
	Dark-throated Thrush *Turdus ruficollis*	w m
R+	Pale-footed Bush Warbler *Cettia pallidipes*	r?
*	Chestnut-crowned Bush Warbler *Cettia major*	w? m?
R	Bright-capped Cisticola *Cisticola exilis*	r
	Fantail Cisticola *Cisticola juncidis*	r
	Graceful Prinia *Prinia gracilis*	r
	Plain Prinia *Prinia inornata*	r
	Ashy Prinia *Prinia socialis*	r
	Grey-breasted Prinia *Prinia hodgsoni*	r
	Yellow-bellied Prinia *Prinia flaviventris*	r
+	Grey-capped Prinia *Prinia cinereocapilla*	r
	Common Tailorbird *Orthotomus sutorius*	r
V	Striated Marsh Warbler *Megalurus palustris*	r
	Black-browed Reed Warbler *Acrocephalus bistrigiceps*	w? m? 5
	Paddyfield Warbler *Acrocephalus agricola*	w
	Blyth's Reed Warbler *Acrocephalus dumetorum*	w m
	Clamorous Reed Warbler *Acrocephalus stentoreus*	w? m?
	Thick-billed Warbler *Acrocephalus aedon*	w? m?
	Golden-spectacled Warbler *Seicercus burkii*	w
	Blyth's Crowned Warbler *Phylloscopus reguloides*	m
	Greenish Warbler *Phylloscopus trochiloides*	w m
	Large-billed Leaf Warbler *Phylloscopus magnirostris*	m
	Yellow-browed Warbler *Phylloscopus inornatus*	m
	Dusky Warbler *Phylloscopus fuscatus*	w
*	Smoky Warbler *Phylloscopus fuligiventer*	w
	Sulphur-bellied Warbler *Phylloscopus griseolus*	w? m?
	Tickell's Warbler *Phylloscopus affinis*	w
	Blue-throated Blue Flycatcher *Cyornis rubeculoides*	m
	Verditer Flycatcher *Muscicapa thalassina*	w m
	Asian Sooty Flycatcher *Muscicapa sibirica*	m
	Asian Brown Flycatcher *Muscicapa latirostris*	m
	Red-breasted Flycatcher *Ficedula parva*	w m
	Grey-headed Flycatcher *Culicicapa ceylonensis*	w m
	White-throated Fantail *Rhipidura albicollis*	r
	Black-naped Monarch *Hypothymis azurea*	r? s?
	Asian Paradise Flycatcher *Terpsiphone paradisi*	s
E	Abbott's Babbler *Trichastoma abbotti*	r
	Striated Babbler *Turdoides earlei*	r
	Jungle Babbler *Turdoides striatus*	r
	Great Tit *Parus major*	r
	Purple Sunbird *Nectarinia asiatica*	r
	Pale-billed Flowerpecker *Dicaeum erythrorhynchos*	r

Oriental White-eye *Zosterops palpebrosa*	r
Black-hooded Oriole *Oriolus xanthornus*	r
Golden Oriole *Oriolus oriolus*	s
Brown Shrike *Lanius cristatus*	w
Bay-backed Shrike *Lanius vittatus*	w? m?
Long-tailed Shrike *Lanius schach*	w
Great Grey Shrike *Lanius excubitor*	r
Black Drongo *Dicrurus macrocercus*	r
Ashy Drongo *Dicrurus leucophaeus*	r
White-bellied Drongo *Dicrurus caerulescens*	r
Bronzed Drongo *Dicrurus aeneus*	w?
Spangled Drongo *Dicrurus hottentottus*	r
Ashy Woodswallow *Artamus fuscus*	r
Red-billed Blue Magpie *Urocissa erythrorhyncha*	r
Rufous Treepie *Dendrocitta vagabunda*	r
House Crow *Corvus splendens*	r
Jungle Crow *Corvus macrorhynchos*	r
Chestnut-tailed Starling *Sturnus malabaricus*	r?
Brahminy Starling *Sturnus pagodarum*	?
Asian Pied Starling *Sturnus contra*	r
Common Mynah *Acridotheres tristis*	r
Bank Mynah *Acridotheres fuscus*	r
Jungle Mynah *Acridotheres fuscus*	r? w? m?
House Sparrow *Passer domesticus*	r
Eurasian Tree Sparrow *Passer montanus*	r
Yellow-throated Sparrow *Petronia xanthocollis*	r
R+ Black-breasted Weaver *Ploceus benghalensis*	r
Baya Weaver *Ploceus philippinus*	r
Red Avadavat *Amandava amandava*	r
Indian Silverbill *Euodice malabarica*	r
Striated Munia *Lonchura striata*	r? m?
Scaly-breasted Munia *Lonchura punctulata*	r
Chestnut Munia *Lonchura malacca*	r
Black-faced Bunting *Emberiza spodocephala*	w
Chestnut-eared Bunting *Emberiza fucata*	w? m?
Yellow-breasted Bunting *Emberiza aureola*	w
Crested Bunting *Melophus lathami*	w? m?

SHIVAPURI WILDLIFE AND WATERSHED RESERVE

	Oriental White-backed Vulture *Gyps bengalensis*	m 4
	Himalayan Griffon Vulture *Gyps himalayensis*	m 4
	Black Kite *Milvus migrans*	r? 4
	Hen Harrier *Circus cyaneus*	w m 4
	Northern Goshawk *Accipiter gentilis*	w 4
	Northern Sparrowhawk *Accipiter nisus*	w 3
	Shikra *Accipiter badius*	r 4
	Common Buzzard *Buteo buteo*	w m 4
	Black Eagle *Ictinaetus malayensis*	r 3
	Steppe Eagle *Aquila rapax nipalensis*	w 3
	Bonelli's Eagle *Hieraaetus fasciatus*	m 5
	Mountain Hawk-Eagle *Spizaetus nipalensis*	r 4
	Common Kestrel *Falco tinnunculus*	r 3
	Common Hill Partridge *Arborophila torqueola*	r 4
	Kalij Pheasant *Lophura leucomelana*	r 4
	Oriental Turtle Dove *Streptopelia orientalis*	r 3
	Ring-necked Parakeet *Psittacula krameri*	?
	Large Hawk-Cuckoo *Hierococcyx sparverioides*	s 3
	Grey-bellied Plaintive Cuckoo *Cacomantis passerinus*	s 4
	Common Cuckoo *Cuculus canorus*	s 3
	Oriental Cuckoo *Cuculus saturatus*	s 3
R	Mountain Scops Owl *Otus spilocephalus*	r 3
	White-throated Needletail *Hirundapus caudacutus*	m 4
	Great Barbet *Megalaima virens*	r 2
	Golden-throated Barbet *Megalaima franklinii*	r 4
+	Darjeeling Pied Woodpecker *Dendrocopos darjellensis*	r 3
+	Crimson-breasted Pied Woodpecker *Dendrocopos cathpharius*	r 4
	Rufous-bellied Pied Woodpecker *Dendrocopos hyperythrus*	r 3
+	Brown-fronted Pied Woodpecker *Dendrocopos auriceps*	r 3
	Barn Swallow *Hirundo rustica*	m 4
	Red-rumped Swallow *Hirundo daurica*	m 4
+	Nepal House-Martin *Delichon nipalensis*	m 4
	Olive-backed Pipit *Anthus hodgsoni*	w 1
	Grey Wagtail *Motacilla cinerea*	r? 1
	Large Cuckoo-shrike *Coracina novaehollandiae*	s 3
	Scarlet Minivet *Pericrocotus flammeus*	r 4
	Long-tailed Minivet *Pericrocotus ethologus*	r 1
E	Grey-chinned Minivet *Pericrocotus solaris*	r? 5
	White-cheeked Bulbul *Pycnonotus leucogenys*	r 1
	Red-vented Bulbul *Pycnonotus cafer*	r 1
	Mountain Bulbul *Hypsipetes mcclellandii*	r 1
	Black Bulbul *Hypsipetes madagascariensis*	r 1
+	Rufous-breasted Accentor *Prunella strophiata*	w 1
	White-tailed Rubythroat *Luscinia pectoralis*	m? 4
	Orange-flanked Bush-Robin *Tarsiger cyanurus*	w 2
+	Golden Bush-Robin *Tarsiger chrysaeus*	w 2
	Asian Magpie-Robin *Copsychus saularis*	r 2
	Blue-capped Redstart *Phoenicurus caeruleocephalus*	w 3
	Hodgson's Redstart *Phoenicurus hodgsoni*	w 2
+	Blue-fronted Redstart *Phoenicurus frontalis*	w 1

	Plumbeous Redstart *Rhyacornis fuliginosus*	r 1
	Common Stonechat *Saxicola torquata*	r 1
	Pied Bushchat *Saxicola caprata*	r 1
	Grey Bushchat *Saxicola ferrea*	r 1
	White-capped Redstart *Chaimarrornis leucocephalus*	w 2
+	Blue-capped Rock-Thrush *Monticola cinclorhyncha*	s 3
	Chestnut-bellied Rock-Thrush *Monticola rufiventris*	r 2
	Blue Whistling Thrush *Myiophoneus caeruleus*	r 1
+	Long-tailed Mountain Thrush *Zoothera dixoni*	w? 3
	Scaly Thrush *Zoothera dauma*	r 4
+	Tickell's Thrush *Turdus unicolor*	s 3
+	White-collared Blackbird *Turdus albocinctus*	w 1
+	Grey-winged Blackbird *Turdus boulboul*	r 3
	Dark-throated Thrush *Turdus ruficollis*	w 3
	Little Forktail *Enicurus scouleri*	? 4
	Spotted Forktail *Enicurus maculatus*	r 2
+	Aberrant Bush Warbler *Cettia flavolivacea*	w 4
I	Yellow-bellied Bush Warbler *Cettia acanthizoides*	r? 4
+	Grey-sided Bush Warbler *Cettia brunnifrons*	w 3
	Striated Prinia *Prinia criniger*	r 1
	Golden-spectacled Warbler *Seicercus burkii*	w 1
+	Grey-hooded Warbler *Seicercus xanthoschistos*	r 1
+	Black-faced Warbler *Abroscopus schisticeps*	r 2
	Blyth's Crowned Warbler *Phylloscopus reguloides*	s 3
	Greenish Warbler *Phylloscopus trochiloides*	m 3
+	Orange-barred Leaf Warbler *Phylloscopus pulcher*	r 1
	Grey-faced Leaf Warbler *Phylloscopus maculipennis*	w 2
	Pallas's Leaf Warbler *Phylloscopus proregulus*	w 1
	Yellow-browed Warbler *Phylloscopus inornatus*	w 2
	Tickell's Warbler *Phylloscopus affinis*	m 3
	Goldcrest *Regulus regulus*	w 4
	Small Niltava *Niltava macgrigoriae*	r? 4
+	Rufous-bellied Niltava *Niltava sundara*	r 4
	Verditer Flycatcher *Muscicapa thalassina*	s 3
	Asian Sooty Flycatcher *Muscicapa sibirica*	s 2
	Slaty-blue Flycatcher *Ficedula tricolor*	m? 4
+	Ultramarine Flycatcher *Ficedula superciliaris*	s 3
	Snowy-browed Flycatcher *Ficedula hyperythra*	s 3
	Orange-gorgetted Flycatcher *Ficedula strophiata*	w 2
	Grey-headed Flycatcher *Culicicapa ceylonensis*	s 3
+	Yellow-bellied Fantail *Rhipidura hypoxantha*	w 3
	White-throated Fantail *Rhipidura albicollis*	r 3
	Rusty-cheeked Scimitar-Babbler *Pomatorhinus erythrogenys*	r 1
	Streak-breasted Scimitar-Babbler *Pomatorhinus ruficollis*	r 2
+	Greater Scaly-breasted Wren-Babbler *Pnoepyga albiventer*	w 2
+	Black-chinned Babbler *Stachyris pyrrhops*	r 1
R	Black-throated Parrotbill *Paradoxornis nipalensis*	r 5
*	Spiny Babbler *Turdoides nipalensis*	r 1
+	White-throated Laughing-thrush *Garrulax albogularis*	r 1
+	Striated Laughing-thrush *Garrulax striatus*	r 1
V+	Rufous-chinned Laughing-thrush *Garrulax rufogularis*	r 4
E+	Grey-sided Laughing-thrush *Garrulax caerulatus*	r 5

	Streaked Laughing-thrush *Garrulax lineatus*	r 3
	Chestnut-crowned Laughing-thrush *Garrulax erythrocephalus*	r 1
	Red-billed Leiothrix *Leiothrix lutea*	r 3
	White-browed Shrike-Babbler *Pteruthius flaviscapis*	r 4
+	Green Shrike-Babbler *Pteruthius xanthochloris*	r 4
	Black-eared Shrike-Babbler *Pteruthius melanotis*	r 4
*	Hoary Barwing *Actinodura nipalensis*	r 1
	Blue-winged Minla *Minla cyanouroptera*	r 3
	Chestnut-tailed Minla *Minla strigula*	r 1
+	Red-tailed Minla *Minla ignotincta*	r 2
	Rufous-winged Fulvetta *Alcippe castaneceps*	r 1
+	White-browed Fulvetta *Alcippe vinipectus*	r 1
+	Nepal Fulvetta *Alcippe nipalensis*	r 2
+	Black-capped Sibia *Heterophasia capistrata*	r 1
+	Whiskered Yuhina *Yuhina flavicollis*	r 1
+	Stripe-throated Yuhina *Yuhina gularis*	r 1
+	Rufous-vented Yuhina *Yuhina occipitalis*	w 3
	White-bellied Yuhina *Yuhina zantholeuca*	r? 4
	Black-throated Tit *Aegithalos concinnus*	r 1
	Yellow-browed Tit *Sylviparus modestus*	r 1
	Great Tit *Parus major*	r? 4
	Green-backed Tit *Parus monticolus*	r 1
	Black-lored Tit *Parus xanthogenys*	r 3
+	White-tailed Nuthatch *Sitta himalayensis*	r 1
	Chestnut-bellied Nuthatch *Sitta castanea*	?
	Wallcreeper *Tichodroma muraria*	w 4
	Brown-throated Treecreeper *Certhia discolor*	r 4
*	Rusty-flanked Treecreeper *Certhia nipalensis*	?
	Green-tailed Sunbird *Aethopyga nipalensis*	r 1
+	Fire-tailed Sunbird *Aethopyga ignicauda*	w 4
	Buff-bellied Flowerpecker *Dicaeum ignipectus*	r 1
	Oriental White-eye *Zosterops palpebrosa*	?
	Maroon Oriole *Oriolus traillii*	r 3
	Long-tailed Shrike *Lanius schach*	r 2
	Ashy Drongo *Dicrurus leucophaeus*	r 3
	Eurasian Jay *Garrulus glandarius*	r 3
	Red-billed Blue Magpie *Urocissa erythrorhyncha*	r 4
	Grey Treepie *Dendrocitta formosae*	r 1
	Yellow-breasted Greenfinch *Carduelis spinoides*	w 3
+	Dark-breasted Rosefinch *Carpodacus nipalensis*	w 2
+	Pink-browed Rosefinch *Carpodacus rhodochrous*	w 3
*	Scarlet Finch *Haematospiza sipahi*	w 3
+	Gold-naped Finch *Pyrrhoplectes epauletta*	w 4
	Brown Bullfinch *Pyrrhula nipalensis*	r 3
+	Red-headed Bullfinch *Pyrrhula erythrocephala*	w 4
	Little Bunting *Emberiza pusilla*	w 3

Confirmation is needed of the continued presence of the following species, all recorded before 1975:

Lammergeier *Gypaetus barbatus*
Eurasian Black Vulture *Aegypius monachus*

	Eurasian Hobby *Falco subbuteo*
R	Oriental Hobby *Falco severus*
+	Speckled Woodpigeon *Columba hodgsonii*
+	Ashy Woodpigeon *Columba pulchricollis*
	Wedge-tailed Green Pigeon *Treron sphenura*
	Little Cuckoo *Cuculus poliocephalus*
E	Forest Eagle Owl *Bubo nipalensis*
	Jungle Nightjar *Caprimulgus indicus*
	Himalayan Swiftlet *Collocalia brevirostris*
	Speckled Piculet *Picumnus innominatus*
	Orange-bellied Leafbird *Chloropsis hardwickii*
	Common Wren *Troglodytes troglodytes*
	Maroon-backed Accentor *Prunella immaculata*
+	Indian Blue Robin *Luscinia brunnea*
R	White-tailed Robin *Cinclidium leucurum*
E*	Purple Cochoa *Cochoa purpurea*
+	Black-backed Forktail *Enicurus immaculatus*
	Slaty-backed Forktail *Enicurus schistaceus*
	Common Tailorbird *Orthotomus sutorius*
+	Large-billed Leaf Warbler *Phylloscopus magnirostris*
	Dusky Warbler *Phylloscopus fuscatus*
R	Ferruginous Flycatcher *Muscicapa ferruginea*
	Asian Brown Flycatcher *Muscicapa latirostris*
V	White-gorgetted Flycatcher *Ficedula monileger*
V+	Slender-billed Scimitar-Babbler *Xiphirhynchus superciliaris*
E+	Blue-winged Laughing-thrush *Garrulax squamatus*
E	Cutia *Cutia nipalensis*

DHORPATAN HUNTING RESERVE

	Black Stork *Ciconia nigra*	m
	Woolly-necked Stork *Ciconia episcopus*	m
	Black Kite *Milvus migrans*	r? s?
	Egyptian Vulture *Neophron percnopterus*	s
	Lammergeier *Gypaetus barbatus*	r
	Himalayan Griffon Vulture *Gyps himalayensis*	r
	Crested Serpent Eagle *Spilornis cheela*	s
	Hen Harrier *Circus cyaneus*	w m
	Pallid Harrier *Circus macrourus*	m
	Northern Sparrowhawk *Accipiter nisus*	s
	Common Buzzard *Buteo buteo*	?
	Common Kestrel *Falco tinnunculus*	r? w? m?
	Chukar Partridge *Alectoris chukar*	r
V+	Satyr Tragopan *Tragopan satyra*	r
	Koklass Pheasant *Pucrasia macrolopha*	r
I*	Cheer Pheasant *Catreus wallichii*	r
	Ibisbill *Ibidorhyncha struthersii*	m
	Eurasian Woodcock *Scolopax rusticola*	s
	Snow Pigeon *Columba leuconota*	r
+	Speckled Woodpigeon *Columba hodgsonii*	s?
	Oriental Turtle Dove *Streptopelia orientalis*	s
	Spotted Dove *Streptopelia chinensis*	s
	Large Hawk-Cuckoo *Hierococcyx sparverioides*	s
	Common Cuckoo *Cuculus canorus*	s
	Oriental Cuckoo *Cuculus saturatus*	s
R	Mountain Scops Owl *Otus spilocephalus*	r
	Asian Barred Owlet *Glaucidium cuculoides*	r
	Tawny Owl *Strix aluco*	r
	Jungle Nightjar *Caprimulgus indicus*	s?
	Alpine Swift *Apus melba*	s?
	Common Kingfisher *Alcedo atthis*	?
	Crested Kingfisher *Ceryle lugubris*	s
	Hoopoe *Upupa epops*	s? m?
	Great Barbet *Megalaima virens*	r? s?
	Scaly-bellied Green Woodpecker *Picus squamatus*	r
+	Himalayan Pied Woodpecker *Dendrocopos himalayensis*	r
	Rufous-bellied Woodpecker *Dendrocopos hyperythrus*	r
+	Brown-fronted Pied Woodpecker *Dendrocopos auriceps*	r
	Oriental Skylark *Alauda gulgula*	s
+	Nepal House-Martin *Delichon nipalensis*	r? s?
	Olive-backed Pipit *Anthus hodgsoni*	r? s?
	Upland Pipit *Anthus sylvanus*	r? s?
	Grey Wagtail *Motacilla cinerea*	s
	White Wagtail *Motacilla alba*	m
	Long-tailed Minivet *Pericrocotous ethologus*	r? s?
	Black Bulbul *Hypsipetes madagascariensis*	r
	Brown Dipper *Cinclus pallasii*	r
+	Indian Blue Robin *Luscinia brunnea*	s
	Orange-flanked Bush-Robin *Tarsiger cyanurus*	r
+	Golden Bush-Robin *Tarsiger chrysaeus*	r?

	Black Redstart *Phoenicurus ochruros*	s
+	Blue-fronted Redstart *Phoenicurus frontalis*	r?
	Plumbeous Redstart *Rhyacornis fuliginosus*	r
+	White-bellied Redstart *Hodgsonius phoenicuroides*	s
	Common Stonechat *Saxicola torquata*	s?
	Grey Bushchat *Saxicola ferrea*	r
	White-capped Redstart *Chaimarrornis leucocephalus*	r? s?
+	Blue-capped Rock-Thrush *Monticola cinclorhyncha*	s
	Blue Whistling Thrush *Myiophoneus caeruleus*	r
*	Pied Ground Thrush *Zoothera wardii*	s
	Orange-headed Ground Thrush *Zoothera citrina*	s
+	White-collared Blackbird *Turdus albocinctus*	r
+	Grey-winged Blackbird *Turdus boulboul*	r? s?
	Dark-throated Thrush *Turdus ruficollis*	w
	Mistle Thrush *Turdus viscivorus*	r
	Little Forktail *Enicurus scouleri*	s
	Brown-flanked Bush Warbler *Cettia fortipes*	s?
+	Grey-sided Bush Warbler *Cettia brunnifrons*	r? s?
	Striated Prinia *Prinia criniger*	r? s?
	Golden-spectacled Warbler *Seicercus burkii*	r? s?
+	Grey-hooded Warbler *Seicercus xanthoschistos*	r
	Blyth's Crowned Warbler *Phylloscopus reguloides*	s
	Western Crowned Warbler *Phylloscopus occipitalis*	s? m?
+	Large-billed Leaf Warbler *Phylloscopus magnirostris*	s
+	Orange-barred Leaf Warbler *Phylloscopus pulcher*	r?
	Grey-faced Leaf Warbler *Phylloscopus maculipennis*	r?
	Yellow-browed Warbler *Phylloscopus inornatus*	r? m?
	Goldcrest *Regulus regulus*	r
	Verditer Flycatcher *Muscicapa thalassina*	s
	Asian Sooty Flycatcher *Muscicapa sibirica*	s
	Slaty-Blue Flycatcher *Ficedula tricolor*	r? s?
+	Ultramarine Flycatcher *Ficedula superciliaris*	s
R	Little Pied Flycatcher *Ficedula westermanni*	s
	Orange-gorgetted Flycatcher *Ficedula strophiata*	r? s?
	Red-breasted Flycatcher *Ficedula parva*	w
	Grey-headed Flycatcher *Culicicapa ceylonensis*	r? s?
+	Yellow-bellied Fantail *Rhipidura hypoxantha*	s
+	White-throated Laughing-thrush *Garrulax albogularis*	r? s?
+	Striated Laughing-thrush *Garrulax striatus*	r
+	Variegated Laughing-thrush *Garrulax variegatus*	r
+	Spotted Laughing-thrush *Garrulax ocellatus*	r
	Streaked Laughing-thrush *Garrulax lineatus*	r
+	Black-faced Laughing-thrush *Garrulax affinis*	r
	Chestnut-tailed Minla *Minla strigula*	r
+	White-browed Fulvetta *Alcippe vinipectus*	r
+	Black-capped Sibia *Heterophasia capistrata*	r
+	Stripe-throated Yuhina *Yuhina gularis*	r
	Black-throated Tit *Aegithalos concinnus*	r
+	Grey-crested Tit *Parus dichrous*	r
+	Rufous-vented Black Tit *Parus rubidiventris*	r
	Spot-winged Black Tit *Parus melanolophus/* Coal Tit *P. ater hybrids*	r

	Green-backed Tit *Parus monticolus*	r
	Black-lored Tit *Parus xanthogenys*	?
+	White-tailed Nuthatch *Sitta himalayensis*	r
	Bar-tailed Treecreeper *Certhia himalayana*	r?
*	Rusty-flanked Treecreeper *Certhia nipalensis*	r
	Common Treecreeper *Certhia familiaris*	r
+	Fire-capped Tit *Cephalopyrus flammiceps*	?
	Green-tailed Sunbird *Aethopyga nipalensis*	r? s?
+	Fire-tailed Sunbird *Aethopyga ignicauda*	r? s?
	Buff-bellied Flowerpecker *Dicaeum ignipectus*	r? s?
	Oriental White-eye *Zosterops palpebrosa*	s
	Long-tailed Shrike *Lanius schach*	s
+	Grey-backed Shrike *Lanius tephronotus*	r? s?
	Ashy Drongo *Dicrurus leucophaeus*	s
+	Lanceolated Jay *Garrulus lanceolatus*	r?
+	Yellow-billed Blue Magpie *Urocissa flavirostris*	r
	Eurasian Nutcracker *Nucifraga caryocatactes*	r
	Red-billed Chough *Pyrrhocorax pyrrhocorax*	r
	Jungle Crow *Corvus macrorhynchos*	r
	Common Mynah *Acridotheres tristis*	r? s?
	House Sparrow *Passer domesticus*	r?
	Cinnamon Sparrow *Passer rutilans*	r?
	Eurasian Tree Sparrow *Passer montanus*	r?
	Scaly-breasted Munia *Lonchura punctulata*	?
	Yellow-breasted Greenfinch *Carduelis spinoides*	s
+	Dark-breasted Rosefinch *Carpodacus nipalensis*	r
	Beautiful Rosefinch *Carpodacus pulcherrimus*	r?
*	Spot-winged Rosefinch *Carpodacus rhodopeplus*	r?
+	Red-headed Bullfinch *Pyrrhula erythrocephala*	r
+	Collared Grosbeak *Mycerobas affinis*	r
	White-winged Grosbeak *Mycerobas carnipes*	r
	Rock Bunting *Emberiza cia*	r
	Chestnut-eared Bunting *Emberiza fucata*	s
	Little Bunting *Emberiza pusilla*	w
	Crested Bunting *Melophus lathami*	s

ANNAPURNA CONSERVATION AREA

Great Crested Grebe *Podiceps cristatus*	v
Great Cormorant *Phalacrocorax carbo*	m 4
Indian Pond Heron *Ardeola grayii*	r 3
Bar-headed Goose *Anser indicus*	m 4
Ruddy Shelduck *Tadorna ferruginea*	m 3
Eurasian Wigeon *Anas penelope*	m 5
Gadwall *Anas strepera*	m 4
Baikal Teal *Anas formosa*	v
Common Teal *Anas crecca*	m 3
Mallard *Anas platyrhynchos*	m b 3
Northern Pintail *Anas acuta*	m 4
Garganey *Anas querquedula*	m 4
Northern Shoveler *Anas clypeata*	m 4
Common Pochard *Aythya ferina*	m 4
Ferruginous Duck *Aythya nyroca*	m 4
Tufted Duck *Aythya fuligula*	m 4
Goosander *Mergus merganser*	w 4
Crested Honey Buzzard *Pernis ptilorhyncus*	r m 3
Black Kite *Milvus migrans*	r m 1
Pallas's Fish Eagle *Haliaeetus leucoryphus*	m 5
Egyptian Vulture *Neophron percnopterus*	s 2
Lammergeier *Gypaetus barbatus*	r b 1
Oriental White-backed Vulture *Gyps bengalensis*	r 3
Long-billed Vulture *Gyps indicus*	r 4
Himalayan Griffon Vulture *Gyps himalayensis*	r 1
Red-headed Vulture *Sarcogyps calvus*	r 2
Eurasian Black Vulture *Aegypius monachus*	w 2
Short-toed Eagle *Circaetus gallicus*	v
Crested Serpent Eagle *Spilornis cheela*	s 1
Eurasian Marsh Harrier *Circus aeruginosus*	m 4
Hen Harrier *Circus cyaneus*	w m 2
Pallid Harrier *Circus macrourus*	m 4
Montagu's Harrier *Circus pygargus*	m 5
Pied Harrier *Circus melanoleucus*	m 5
Northern Goshawk *Accipiter gentilis*	r 3
Besra *Accipiter virgatus*	r 4
Northern Sparrowhawk *Accipiter nisus*	r 2
Crested Goshawk *Accipiter trivirgatus*	r 4
Common Buzzard *Buteo buteo*	w m 2
Long-legged Buzzard *Buteo rufinus*	w? m 2
Upland Buzzard *Buteo hemilasius*	?
Black Eagle *Ictinaetus malayensis*	r 2
Greater Spotted Eagle *Aquila clanga*	m 3
Steppe Eagle *Aquila rapax nipalensis*	w m 1
Imperial Eagle *Aquila heliaca*	m 4
Golden Eagle *Aquila chrysaetos*	r 4
Booted Eagle *Hieraaetus pennatus*	w? m b 4
Bonelli's Eagle *Hieraaetus fasciatus*	r 3
Mountain Hawk-Eagle *Spizaetus nipalensis*	r 3
Osprey *Pandion haliaetus*	m 4

	Lesser Kestrel *Falco naumanni*	m 4
	Common Kestrel *Falco tinnunculus*	r m w? 1
	Amur Falcon *Falco amurensis*	m 5
	Merlin *Falco columbarius*	v
	Eurasian Hobby *Falco subbuteo*	r? w? b 4
	Saker *Falco cherrug*	w 5
	Peregrine *Falco peregrinus*	r 2
	Barbary Falcon *Falco pelegrinoides*	? 5
+	Snow Partridge *Lerwa lerwa*	r?
	Tibetan Snowcock *Tetraogallus tibetanus*	r 2
	Himalayan Snowcock *Tetraogallus himalayensis*	r?
	Chukar Partridge *Alectoris chukar*	r 2
	Black Francolin *Francolinus francolinus*	s 1
	Tibetan Partridge *Perdix hodgsoniae*	r?
	Common Hill Partridge *Arborophila torqueola*	r 2
E	Rufous-throated Hill Partridge *Arborophila rufogularis*	r 5
+	Blood Pheasant *Ithaginis cruentus*	r 2
V+	Satyr Tragopan *Tragopan satyra*	r 4
	Koklass Pheasant *Pucrasia macrolophus*	r 2
+	Himalayan Monal *Lophophorus impejanus*	r 2
	Kalij Pheasant *Lophura leucomelana*	r 2
I*	Cheer Pheasant *Catreus wallichii*	r?
	Common Moorhen *Gallinula chloropus*	m 5
	Common Coot *Fulica atra*	v
	Common Crane *Grus grus*	m 5
	Demoiselle Crane *Anthropoides virgo*	m 1
	Black-winged Stilt *Himantopus himantopus*	v
	Little Ringed Plover *Charadrius dubius*	m 5
	Lesser Golden Plover *Pluvialis dominica*	
	(Pacific Golden Plover *P. fulva*)	v
	Northern Lapwing *Vanellus vanellus*	v
	Temminck's Stint *Calidris temminckii*	m 3
	Solitary Snipe *Gallinago solitaria*	w 3
I*	Wood Snipe *Gallinago nemoricola*	? 5
	Eurasian Woodcock *Scolopax rusticola*	r 3
	Common Greenshank *Tringa nebularia*	m 5
	Green Sandpiper *Tringa ochropus*	w m 2
	Wood Sandpiper *Tringa glareola*	m 3
	Common Sandpiper *Actitis hypoleucos*	m 2
	Ruddy Turnstone *Arenaria interpres*	v
	Brown-headed Gull *Larus brunnicephalus*	v
	Rock Pigeon *Columba livia*	r 1
	Hill Pigeon *Columba rupestris*	r 2
	Common Woodpigeon *Columba palumbus*	w 5
+	Speckled Woodpigeon *Columba hodgsonii*	r 3
+	Ashy Woodpigeon *Columba pulchricollis*	r 3
	Oriental Turtle Dove *Streptopelia orientalis*	r 1
	Spotted Dove *Streptopelia chinensis*	r 1
V	Barred Cuckoo-Dove *Macropygia unchall*	r 4
	Wedge-tailed Green Pigeon *Treron sphenura*	r 4
+	Slaty-headed Parakeet *Psittacula himalayana*	r 1
	Large Hawk-Cuckoo *Hierococcyx sparverioides*	s 2

	Indian Cuckoo *Cuculus micropterus*	s 1
	Common Cuckoo *Cuculus canorus*	s 1
	Oriental Cuckoo *Cuculus saturatus*	s 1
	Lesser Cuckoo *Cuculus poliocephalus*	s 4
	Green-billed Malkoha *Phaenicophaeus tristis*	r 4
	Oriental Scops Owl *Otus sunia*	v
R	Mountain Scops Owl *Otus spilocephalus*	r 2
	Northern Eagle Owl *Bubo bubo*	r 4
	Collared Owlet *Glaucidium brodiei*	r 2
	Jungle Owlet *Glaucidium radiatum*	r? 5
	Asian Barred Owlet *Glaucidium cuculoides*	r 1
	Northern Little Owl *Athene noctua*	r 5
	Spotted Little Owl *Athene brama*	r 3
	Tawny Owl *Strix aluco*	r 4
	Short-eared Owl *Asio flammeus*	w? m? 5
	Jungle Nightjar *Caprimulgus indicus*	r 2
	Himalayan Swiftlet *Collocalia brevirostris*	r 2
	White-throated Needletail *Hirundapus caudacutus*	?
	Common Swift *Apus apus*	s 2
	Pacific Swift *Apus pacificus*	s 2
	Alpine Swift *Apus melba*	? 2
	Little Swift *Apus affinis*	r 1
E	Red-headed Trogon *Harpactes erythrocephalus*	r 5
	White-breasted Kingfisher *Halcyon smyrnensis*	r? 5
	Common Kingfisher *Alcedo atthis*	r 3
	Crested Kingfisher *Ceryle lugubris*	r 3
	Green Bee-eater *Merops orientalis*	m 5
	Indian Roller *Coracias benghalensis*	r? 5
	Hoopoe *Upupa epops*	r s m 1
	Great Barbet *Megalaima virens*	r 1
	Golden-throated Barbet *Megalaima franklinii*	r 3
	Blue-throated Barbet *Megalaima asiatica*	r 2
R*	Orange-rumped Honeyguide *Indicator xanthonotus*	r 4
	Eurasian Wryneck *Jynx torquilla*	m 3
	Speckled Piculet *Picumnus innominatus*	r 3
	Rufous Woodpecker *Celeus brachyurus*	r? 5
	Lesser Yellow-naped Woodpecker *Picus chlorolophus*	r 3
	Greater Yellow-naped Woodpecker *Picus flavinucha*	r 3
	Grey-headed Woodpecker *Picus canus*	r 2
	Scaly-bellied Green Woodpecker *Picus squamatus*	r 2
	Greater Golden-backed Woodpecker *Chrysocolaptes lucidus*	r 5
V	Bay Woodpecker *Blythipicus pyrrhotis*	r 5
+	Darjeeling Pied Woodpecker *Dendrocopos darjellensis*	r 2
+	Crimson-breasted Pied Woodpecker *Dendrocopos cathpharius*	r 2
	Rufous-bellied Pied Woodpecker *Dendrocopos hyperythrus*	r 2
+	Brown-fronted Pied Woodpecker *Dendrocopos auriceps*	r 2
	Fulvous-breasted Pied Woodpecker *Dendrocopos macei*	r?
	Greater Short-toed Lark *Calandrella brachydactyla*	m 1
	Hume's Short-toed Lark *Calandrella acutirostris*	m 4
	Oriental Skylark *Alauda gulgula*	r w 2
	Horned Lark *Eremophila alpestris*	w 5
	Brown-throated Sand Martin *Riparia paludicola*	r 1

	Collared Sand Martin *Riparia riparia*	m 5
	Crag Martin *Ptyonoprogne rupestris*	r 2
	Barn Swallow *Hirundo rustica*	r s b 1
	Red-rumped Swallow *Hirundo daurica*	r s b 1
+	Nepal House-Martin *Delichon nipalensis*	r 2
	Asian House-Martin *Delichon dasypus*	?
	Richard's Pipit *Anthus novaeseelandiae*	w m 3
	Olive-backed Pipit *Anthus hodgsoni*	r 1
	Tree Pipit *Anthus trivialis*	m 5
	Red-throated Pipit *Anthus cervinus*	v
	Rosy Pipit *Anthus roseatus*	r m b 2
	Water Pipit *Anthus spinoletta*	m 5
	Upland Pipit *Anthus sylvanus*	r 2
	Yellow Wagtail *Motacilla flava*	m 3
	Citrine Wagtail *Motacilla citreola*	m 4
	Grey Wagtail *Motacilla cinerea*	r 1
	White Wagtail *Motacilla alba*	r? w m 1
	White-browed Wagtail *Motacilla maderaspatensis*	r 5
	Large Cuckoo-shrike *Coracina novaehollandiae*	r 1
	Bar-winged Flycatcher-shrike *Hemipus picatus*	r? 3
	Scarlet Minivet *Pericrocotus flammeus*	r 1
R	Short-billed Minivet *Pericrocotus brevirostris*	r 5
	Long-tailed Minivet *Pericrocotus ethologus*	r 1
E	Grey-chinned Minivet *Pericrocotus solaris*	r 5
+	Striated Bulbul *Pycnonotus striatus*	r 3
	White-cheeked Bulbul *Pycnonotus leucogenys*	r 1
	Red-vented Bulbul *Pycnonotus cafer*	r 1
	Mountain Bulbul *Hypsipetes mcclellandii*	r 2
	Black Bulbul *Hypsipetes madagascariensis*	r 1
	Orange-bellied Leafbird *Chloropsis hardwickii*	r 3
	White-breasted Dipper *Cinclus cinclus*	?
	Brown Dipper *Cinclus pallasii*	r 1
	Northern Wren *Troglodytes troglodytes*	r 2
	Maroon-backed Accentor *Prunella immaculata*	w 3
+	Rufous-breasted Accentor *Prunella strophiata*	r 2
	Brown Accentor *Prunella fulvescens*	r? w 2
	Black-throated Accentor *Prunella atrogularis*	w 5
+	Robin Accentor *Prunella rubeculoides*	w 2
	Altai Accentor *Prunella himalayana*	w 2
	Alpine Accentor *Prunella collaris*	w 2
R*	Gould's Shortwing *Brachypteryx stellata*	? 5
R	White-browed Shortwing *Brachypteryx montana*	? 5
	Bluethroat *Luscinia svecica*	m 4
	White-tailed Rubythroat *Luscinia pectoralis*	s 3
+	Indian Blue Robin *Luscinia brunnea*	s 2
	Orange-flanked Bush-Robin *Tarsiger cyanurus*	r 1
+	Golden Bush-Robin *Tarsiger chrysaeus*	r 3
+	White-browed Bush-Robin *Tarsiger indicus*	r 3
*	Rufous-breasted Bush-Robin *Tarsiger hyperythrus*	r? 3
	Asian Magpie-Robin *Copsychus saularis*	r 3
	Rufous-backed Redstart *Phoenicurus erythronotus*	w 3
	Blue-capped Redstart *Phoenicurus caeruleocephalus*	r 2

	Black Redstart *Phoenicurus ochruros*	s b 2
	Hodgson's Redstart *Phoenicurus hodgsoni*	w 1
+	Blue-fronted Redstart *Phoenicurus frontalis*	r 1
+	White-throated Redstart *Phoenicurus schisticeps*	r 3
	Güldenstadt's Redstart *Phoenicurus erythrogaster*	w 2
	Plumbeous Redstart *Rhyacornis fuliginosus*	r 1
+	White-bellied Redstart *Hodgsonius phoenicuroides*	s b 3
R	White-tailed Robin *Cinclidium leucurum*	r 4
+	Grandala *Grandala coelicolor*	r? w 2
	Common Stonechat *Saxicola torquata*	r m w 1
	Pied Bushchat *Saxicola caprata*	r 2
	Grey Bushchat *Saxicola ferrea*	r 2
	Isabelline Wheatear *Oenanthe isabellina*	m 5
	Desert Wheatear *Oenanthe deserti*	s m 5
	White-capped Redstart *Chaimarrornis leucocephalus*	r 1
+	Blue-capped Rock-Thrush *Monticola cinclorhyncha*	s 3
	Chestnut-bellied Rock-Thrush *Monticola rufiventris*	r 2
	Blue Rock-Thrush *Monticola solitarius*	r 2
	Blue Whistling Thrush *Myiophoneus caeruleus*	r 1
+	Plain-backed Mountain Thrush *Zoothera mollissima*	r 2
+	Long-tailed Mountain Thrush *Zoothera dixoni*	r 3
	Scaly Thrush *Zoothera dauma*	r? 3
R*	Long-billed Thrush *Zoothera monticola*	r? 4
*	Pied Ground Thrush *Zoothera wardii*	s 4
	Orange-headed Ground Thrush *Zoothera citrina*	s 4
+	Tickell's Thrush *Turdus unicolor*	s 3
+	White-collared Blackbird *Turdus albocinctus*	r 2
+	Grey-winged Blackbird *Turdus boulboul*	r 2
	Eurasian Blackbird *Turdus merula*	m 5
	Chestnut Thrush *Turdus rubrocanus*	w 4
	Rufous-tailed Thrush *Turdus naumanni*	w 5
	Dark-throated Thrush *Turdus ruficollis*	w 1
	Mistle Thrush *Turdus viscivorus*	r 2
	Little Forktail *Enicurus scouleri*	r 2
+	Black-backed Forktail *Enicurus immaculatus*	r 4
	Slaty-backed Forktail *Enicurus schistaceus*	r 2
	Spotted Forktail *Enicurus maculatus*	r 1
+	Chestnut-headed Tesia *Tesia castaneocoronata*	r 2
	Grey-bellied Tesia *Tesia cyaniventer*	r 3
*	Chestnut-crowned Bush Warbler *Cettia major*	s b 5
+	Aberrant Bush Warbler *Cettia flavolivacea*	r 1
I	Yellow-bellied Bush Warbler *Cettia acanthizoides*	s 5
+	Grey-sided Bush Warbler *Cettia brunnifrons*	r 1
	Spotted Bush Warbler *Bradypterus thoracicus*	s b 5
	Striated Prinia *Prinia criniger*	r 1
	Common Tailorbird *Orthotomus sutorius*	r 1
	Blyth's Reed Warbler *Acrocephalus dumetorum*	w 5
	Booted Warbler *Hippolais caligata*	v
	Lesser Whitethroat *Sylvia curruca*	w? m 5
	Golden-spectacled Warbler *Seicercus burkii*	r 1
E+	Grey-cheeked Warbler *Seicercus poliogenys*	r? 5
	Chestnut-crowned Warbler *Seicercus castaniceps*	r 3

+	Grey-hooded Warbler *Seicercus xanthoschistos*	r b 1
+	Black-faced Warbler *Abroscopus schisticeps*	r 2
	Blyth's Crowned Warbler *Phylloscopus reguloides*	r 1
	Western Crowned Warbler *Phylloscopus occipitalis*	m 5
	Greenish Warbler *Phylloscopus trochiloides*	s w m 2
+	Large-billed Leaf Warbler *Phylloscopus magnirostris*	s 4
+	Orange-barred Leaf Warbler *Phylloscopus pulcher*	r 1
	Grey-faced Leaf Warbler *Phylloscopus maculipennis*	r 1
	Pallas's Leaf Warbler *Phylloscopus proregulus*	r 1
	Yellow-browed Warbler *Phylloscopus inornatus*	w m 1
*	Smoky Warbler *Phylloscopus fuligiventer*	s 5
	Tickell's Warbler *Phylloscopus affinis*	r 2
	Chiffchaff *Phylloscopus collybita*	w m 2
	Goldcrest *Regulus regulus*	r 2
	Stoliczka's Tit-Warbler *Leptopoecile sophiae*	r b 3
V	Large Niltava *Niltava grandis*	r 4
	Small Niltava *Niltava macgrigoriae*	s 2
+	Rufous-bellied Niltava *Niltava sundara*	r 1
V	Hill Blue Flycatcher *Cyornis banyumas*	m 5
V	Pygmy Blue Flycatcher *Muscicapella hodgsoni*	r 5
	Verditer Flycatcher *Muscicapa thalassina*	s 1
R	Ferruginous Flycatcher *Muscicapa ferruginea*	s 5
	Asian Sooty Flycatcher *Muscicapa sibirica*	s 2
+	Rufous-tailed Flycatcher *Muscicapa ruficauda*	s 4
	Asian Brown Flycatcher *Muscicapa latirostris*	s?
	Slaty-blue Flycatcher *Ficedula tricolor*	r 1
+	Ultramarine Flycatcher *Ficedula superciliaris*	s 1
R	Little Pied Flycatcher *Ficedula westermanni*	s 4
+	Slaty-backed Flycatcher *Ficedula hodgsonii*	m 5
	Snowy-browed Flycatcher *Ficedula hyperythra*	s 3
V	White-gorgetted Flycatcher *Ficedula moniliger*	r? 5
	Orange-gorgetted Flycatcher *Ficedula strophiata*	r 1
	Red-breasted Flycatcher *Ficedula parva*	w m 2
	Grey-headed Flycatcher *Culicicapa ceylonensis*	r 4 s 1
+	Yellow-bellied Fantail *Rhipidura hypoxantha*	r 1
	White-throated Fantail *Rhipidura albicollis*	r 3
	Rusty-cheeked Scimitar-Babbler *Pomatorhinus erythrogenys*	r 1
	White-browed Scimitar-Babbler *Pomatorhinus schisticeps*	r 3
	Streak-breasted Scimitar-Babbler *Pomatorhinus ruficollis*	r 2
V+	Slender-billed Scimitar-Babbler *Xiphirhynchus superciliaris*	r 5
+	Greater Scaly-breasted Wren-Babbler *Pnoepyga albiventer*	r 2
	Lesser Scaly-breasted Wren-Babbler *Pnoepyga pusilla*	r 3
+	Black-chinned Babbler *Stachyris pyrrhops*	r 2
E	Golden Babbler *Stachyris chrysaea*	r 5
	Grey-throated Babbler *Stachyris nigriceps*	r? 5
V*	Great Parrotbill *Conostoma aemodium*	r 4
V*	Brown Parrotbill *Paradoxornis unicolor*	r 4
V*	Fulvous Parrotbill *Paradoxornis fulvifrons*	r 4
R	Black-throated Parrotbill *Paradoxornis nipalensis*	r 4
*	Spiny Babbler *Turdoides nipalensis*	r 3
+	White-throated Laughing-thrush *Garrulax albogularis*	r 1
	White-crested Laughing-thrush *Garrulax leucolophus*	r 1

+	Striated Laughing-thrush *Garrulax striatus*	r 1
+	Variegated Laughing-thrush *Garrulax variegatus*	r 1
V+	Rufous-chinned Laughing-thrush *Garrulax rufogularis*	r 3
+	Spotted Laughing-thrush *Garrulax ocellatus*	r 2
E+	Grey-sided Laughing-thrush *Garrulax caerulatus*	r? 5
	Streaked Laughing-thrush *Garrulax lineatus*	r 1
E+	Blue-winged Laughing-thrush *Garrulax squamatus*	r 5
V+	Scaly Laughing-thrush *Garrulax subunicolor*	r 3
+	Black-faced Laughing-thrush *Garrulax affinis*	r 1
	Chestnut-crowned Laughing-thrush *Garrulax erythrocephalus*	r 1
	Red-billed Leiothrix *Leiothrix lutea*	r 2
R*	Fire-tailed Myzornis *Myzornis pyrrhoura*	r? 5
E	Cutia *Cutia nipalensis*	r 5
E*	Black-headed Shrike-Babbler *Pteruthius rufiventer*	r 5
	White-browed Shrike-Babbler *Pteruthius flaviscapis*	r 2
+	Green Shrike-Babbler *Pteruthius xanthochloris*	r 3
	Black-eared Shrike-Babbler *Pteruthius melanotis*	r 3
*	Hoary Barwing *Actinodura nipalensis*	r 2
	Blue-winged Minla *Minla cyanouroptera*	r 2
	Chestnut-tailed Minla *Minla strigula*	r 1
+	Red-tailed Minla *Minla ignotincta*	r 4
V*	Golden-breasted Fulvetta *Alcippe chrysotis*	r 2
	Rufous-winged Fulvetta *Alcippe castaneceps*	r 1
+	White-browed Fulvetta *Alcippe vinipectus*	r 1
+	Nepal Fulvetta *Alcippe nipalensis*	r 4
+	Black-capped Sibia *Heterophasia capistrata*	r 1
+	Whiskered Yuhina *Yuhina flavicollis*	r 1
+	Stripe-throated Yuhina *Yuhina gularis*	r 1
+	Rufous-vented Yuhina *Yuhina occipitalis*	r 1
	White-bellied Yuhina *Yuhina zantholeuca*	r 3
+	Black-browed Tit *Aegithalos iouschistos*	r 4
	White-throated Tit *Aegithalos niveogularis*	? 5
	Black-throated Tit *Aegithalos concinnus*	r 1
	Yellow-browed Tit *Sylviparus modestus*	r 1
+	Grey-crested Tit *Parus dichrous*	r 1
	Rufous-naped Black Tit *Parus rufonuchalis*	r? b 3
+	Rufous-vented Black Tit *Parus rubidiventris*	r 1
	Coal Tit *Parus ater*	r 1
	Great Tit *Parus major*	r 2
	Green-backed Tit *Parus monticolus*	r 1
	Black-lored Tit *Parus xanthogenys*	r 1
	Velvet-fronted Nuthatch *Sitta frontalis*	r 2
+	White-tailed Nuthatch *Sitta himalayensis*	r 1
	Chestnut-bellied Nuthatch *Sitta castanea*	r 1
	Wallcreeper *Tichodroma muraria*	w 1
	Brown-throated Treecreeper *Certhia discolor*	r 4
	Bar-tailed Treecreeper *Certhia himalayana*	r 2
*	Rusty-flanked Treecreeper *Certhia nipalensis*	r 2
	Common Treecreeper *Certhia familiaris*	r 2
+	Fire-capped Tit *Cephalopyrus flammiceps*	? 4
	Purple Sunbird *Nectarinia asiatica*	s 3
	Mrs Gould's Sunbird *Aethopyga gouldiae*	r 4

	Green-tailed Sunbird *Aethopyga nipalensis*	r 1
	Black-throated Sunbird *Aethopyga saturata*	r 3
	Crimson Sunbird *Aethopyga siparaja*	r 3
+	Fire-tailed Sunbird *Aethopyga ignicauda*	r 2
R*	Yellow-bellied Flowerpecker *Dicaeum melanoxanthum*	r? 4
	Buff-bellied Flowerpecker *Dicaeum ignipectus*	r 1
	Oriental White-eye *Zosterops palpebrosa*	r 1
	Maroon Oriole *Oriolus traillii*	r 2
	Eurasian Golden Oriole *Oriolus oriolus*	s 3
	Brown Shrike *Lanius cristatus*	w m 4
	Isabelline Shrike *Lanius isabellinus*	v
	Bay-backed Shrike *Lanius vittatus*	v
	Long-tailed Shrike *Lanius schach*	r 1
+	Grey-backed Shrike *Lanius tephronotus*	r 2
	Black Drongo *Dicrurus macrocercus*	r 1
	Ashy Drongo *Dicrurus leucophaeus*	r 1
	Bronzed Drongo *Dicrurus aeneus*	s 3
	Lesser Racket-tailed Drongo *Dicrurus remifer*	r 3
	Spangled Drongo *Dicrurus hottentottus*	r 3
	Ashy Woodswallow *Artamus fuscus*	m 5
	Eurasian Jay *Garrulus glandarius*	r 4
+	Lanceolated Jay *Garrulus lanceolatus*	r 4
+	Yellow-billed Blue Magpie *Urocissa flavirostris*	r 1
	Red-billed Blue Magpie *Urocissa erythrorhyncha*	r 1
	Green Magpie *Cissa chinensis*	r 2
	Grey Treepie *Dendrocitta formosae*	r 1
	Hume's Ground Jay *Pseudopodoces humilis*	?
	Eurasian Nutcracker *Nucifraga caryocatactes*	r 1
	Alpine Chough *Pyrrhocorax graculus*	r 1
	Red-billed Chough *Pyrrhocorax pyrrhocorax*	r 1
	Jungle Crow *Corvus macrorhynchos*	r 1
	Common Raven *Corvus corax*	r 3
	Common Starling *Sturnus vulgaris*	w m 5
	Common Mynah *Acridotheres tristis*	r 1
	House Sparrow *Passer domesticus*	r 1
	Cinnamon Sparrow *Passer rutilans*	r 2
	Eurasian Tree Sparrow *Passer montanus*	r 1
	Red-necked Snowfinch *Montifringilla ruficollis*	v
	Tibetan Snowfinch *Montifringilla adamsi*	r?
	Baya Weaver *Ploceus philippinus*	s?
	Striated Munia *Lonchura striata*	r 3
	Scaly-breasted Munia *Lonchura punctulata*	r 4
	Chestnut Munia *Lonchura malacca*	? 5
	Common Chaffinch *Fringilla coelebs*	w 3
	Brambling *Fringilla montifringilla*	w 4
	Red-fronted Serin *Serinus pusillus*	r 2
	Tibetan Serin *Serinus thibetanus*	w 4
+	Red-browed Finch *Callacanthis burtoni*	r? w 4
	Yellow-breasted Greenfinch *Carduelis spinoides*	r 1
	Eurasian Goldfinch *Carduelis carduelis*	r 4
	Twite *Carduelis flavirostris*	r 4
	Common Crossbill *Loxia curvirostra*	? 3

	Plain Mountain-Finch *Leucosticte nemoricola*	r 1
	Brandt's Mountain-Finch *Leucosticte brandti*	? 3
	Mongolian Finch *Bucanetes mongolicus*	v
*	Crimson Rosefinch *Carpodacus rubescens*	w 5
+	Dark-breasted Rosefinch *Carpodacus nipalensis*	r 2
	Common Rosefinch *Carpodacus erythrinus*	r 2
	Beautiful Rosefinch *Carpodacus pulcherrimus*	r 1
+	Pink-browed Rosefinch *Carpodacus rhodochrous*	r 2
*	Vinaceous Rosefinch *Carpodacus vinaceus*	? 5
*	Spot-winged Rosefinch *Carpodacus rhodopeplus*	r 2
+	White-browed Rosefinch *Carpodacus thura*	r 2
+	Crimson-eared Rosefinch *Carpodacus rubicilloides*	w 2
	Spot-crowned Rosefinch *Carpodacus rubicilla*	r 3
	Red-breasted Rosefinch *Carpodacus puniceus*	r? w 4
*	Crimson-browed Finch *Propyrrhula subhimachala*	r? w 3
V*	Scarlet Finch *Haematospiza sipahi*	r 4
+	Gold-naped Finch *Pyrrhoplectes epauletta*	r? 5
	Brown Bullfinch *Pyrrhula nipalensis*	r 4
+	Red-headed Bullfinch *Pyrrhula erythrocephala*	r 2
+	Collared Grosbeak *Mycerobas affinis*	r 2
*	Spot-winged Grosbeak *Mycerobas melanozanthos*	r? 5
	White-winged Grosbeak *Mycerobas carnipes*	r 2
	Pine Bunting *Emberiza leucocephalos*	w 2
	Yellowhammer *Emberiza citrinella*	w 5
	Rock Bunting *Emberiza cia*	r 1
	Chestnut-eared Bunting *Emberiza fucata*	? 5
	Rustic Bunting *Emberiza rustica*	v
	Little Bunting *Emberiza pusilla*	w 2
	Crested Bunting *Melophus lathami*	r 2

MAI VALLEY WATERSHED (not tropical zone)

	Indian Pond Heron *Ardeola grayii*	r
	Little Egret *Egretta garzetta*	r
	Black Stork *Ciconia nigra*	m 5
	Woolly-necked Stork *Ciconia episcopus*	m 5
	Bar-headed Goose *Anser indicus*	m 5
	Ruddy Shelduck *Tadorna ferruginea*	m 5
	Tufted Duck *Aythya fuligula*	m 5
	Crested Honey Buzzard *Pernis ptilorhyncus*	r
	Black Kite *Milvus migrans*	r 2
	Lammergeier *Gypaetus barbatus*	m 5
	Oriental White-backed Vulture *Gyps bengalensis*	r 1
	Eurasian Griffon Vulture *Gyps fulvus*	r 4
	Red-headed Vulture *Sarcogyps calvus*	r 4
	Crested Serpent Eagle *Spilornis cheela*	s 1
	Hen Harrier *Circus cyaneus*	w m 2
	Northern Goshawk *Accipiter gentilis*	r? 3
	Besra *Accipiter virgatus*	r 4
	Northern Sparrowhawk *Accipiter nisus*	r w 2
	Crested Goshawk *Accipiter trivirgatus*	?
	Shikra *Accipiter badius*	r 2
	Common Buzzard *Buteo buteo*	w m 2
	Upland Buzzard *Buteo hemilasius*	w? m?
	Black Eagle *Ictinaetus malayensis*	r 2
	Steppe Eagle *Aquila rapax nipalensis*	w m 2
	Mountain Hawk-Eagle *Spizaetus nipalensis*	r 2
	Red-thighed Falconet *Microhierax caerulescens*	r 3
	Common Kestrel *Falco tinnunculus*	r? w m 2
	Common Hill Partridge *Arborophila torqueola*	r 2
	Kalij Pheasant *Lophura leucomelana*	r 2
	Solitary Snipe *Gallinago solitaria*	w? m? 4
I*	Wood Snipe *Gallinago nemoricola*	?
	(only recorded by Stevens 1925)	
	Eurasian Woodcock *Scolopax rusticola*	s? m? 5
	Green Sandpiper *Tringa ochropus*	m 5
+	Speckled Woodpigeon *Columba hodgsonii*	r?
+	Ashy Woodpigeon *Columba pulchricollis*	r
	Oriental Turtle Dove *Streptopelia orientalis*	r 1
	Spotted Dove *Streptopelia chinensis*	r 1
	Wedge-tailed Green Pigeon *Treron sphenura*	r
+	Slaty-headed Parakeet *Psittacula himalayana*	r
	Common Hawk-Cuckoo *Hierococcyx varius*	r 2
	Large Hawk-Cuckoo *Hierococcyx sparverioides*	s 2
	Indian Cuckoo *Cuculus micropterus*	s
	Common Cuckoo *Cuculus canorus*	s 1
	Oriental Cuckoo *Cuculus saturatus*	s
	Lesser Cuckoo *Cuculus poliocephalus*	?
	(only recorded by Stevens 1925)	
	Green-billed Malkoha *Phaenicophaeus tristis*	r? s
E	Forest Eagle Owl *Bubo nipalensis*	?
	(proved breeding, only recorded by Stevens 1925)	

	Collared Owlet *Glaucidium brodiei*	r
	Asian Barred Owlet *Glaucidium cuculoides*	r 1
V	Brown Wood Owl *Strix leptogrammica*	r 5
	Short-eared Owl *Asio flammeus*	m 5
	Jungle Nightjar *Caprimulgus indicus*	r? s?
	Himalayan Swiftlet *Collocalia brevirostris*	r 2
	White-throated Needletail *Hirundapus caudacutus*	? 5
	Pacific Swift *Apus pacificus*	?
	Alpine Swift *Apus melba*	?
	Little Swift *Apus affinis*	s? 1
	White-breasted Kingfisher *Halcyon smyrnensis*	r 2
	Common Kingfisher *Alcedo atthis*	r
	Crested Kingfisher *Ceryle lugubris*	r
	Chestnut-headed Bee-eater *Merops leschenaulti*	s
	Indian Roller *Coracias benghalensis*	r 1
	Hoopoe *Upupa epops*	m
	Great Barbet *Megalaima virens*	r 1
	Blue-throated Barbet *Megalaima asiatica*	r 2
	Speckled Piculet *Picumnus innominatus*	r
R	White-browed Piculet *Sasia ochracea*	r 4
	Lesser Yellow-naped Woodpecker *Picus chlorolophus*	r
	Greater Yellow-naped Woodpecker *Picus flavinucha*	r
	Grey-headed Woodpecker *Picus canus*	r
V	Bay Woodpecker *Blythipicus pyrrhotis*	r 4
+	Darjeeling Pied Woodpecker *Dendrocopos darjellensis*	r 1
+	Crimson-breasted Pied Woodpecker *Dendrocopos cathpharius*	r
+	Rufous-bellied Pied Woodpecker *Dendrocopos hyperythrus* (only recorded by Stevens 1925)	?
	Fulvous-breasted Pied Woodpecker *Dendrocopos macei*	r
	Grey-capped Pygmy Woodpecker *Dendrocopos canicapillus*	r
E	Long-tailed Broadbill *Psarisomus dalhousiae*	r 5
	Barn Swallow *Hirundo rustica*	r? s 1
	Red-rumped Swallow *Hirundo daurica*	r? s 1
+	Nepal House-Martin *Delichon nipalensis*	r 2
	Richard's Pipit *Anthus novaeseelandiae*	r w? m?
	Olive-backed Pipit *Anthus hodgsoni*	w 1
	Upland Pipit *Anthus sylvanus*	r
	Grey Wagtail *Motacilla cinerea*	r? s
	White Wagtail *Motacilla alba*	w m
	Bar-winged Flycatcher-shrike *Hemipus picatus*	r
	Black-winged Cuckoo-shrike *Coracina melaschistos*	s
	Large Cuckoo-shrike *Coracina novaehollandiae*	r
	Scarlet Minivet *Pericrocotus flammeus*	r
	Short-billed Minivet *Pericrocotus brevirostris*	r b
	Long-tailed Minivet *Pericrocotus ethologus*	r 1
E	Grey-chinned Minivet *Pericrocotus solaris*	r? w? 5
+	Striated Bulbul *Pycnonotus striatus*	r
	White-cheeked Bulbul *Pycnonotus leucogenys*	r 1
	Red-vented Bulbul *Pycnonotus cafer*	r 1
	Mountain Bulbul *Hypsipetes mcclellandii*	r?
	Ashy Bulbul *Hypsipetes flavalus*	r?
	Black Bulbul *Hypsipetes madagascariensis*	r 1

	Common Iora *Aegithina tiphia*	r
	Orange-bellied Leafbird *Chloropsis hardwickii*	r
	Brown Dipper *Cinclus pallasii*	r
	Northern Wren *Troglodytes troglodytes*	w
	Maroon-backed Accentor *Prunella immaculata*	w 2
+	Rufous-breasted Accentor *Prunella strophiata*	w 2
R*	Gould's Shortwing *Brachypteryx stellata*	w? 5
R	White-browed Shortwing *Brachypteryx montana*	r? s 4
E	Lesser Shortwing *Brachypteryx leucophrys*	r? w? 5
	White-tailed Rubythroat *Luscinia pectoralis*	w? m?
+	Indian Blue Robin *Luscinia brunnea*	s
	Orange-flanked Bush-Robin *Tarsiger cyanurus*	w 1
+	Golden Bush-Robin *Tarsiger chrysaeus*	w 2
+	White-browed Bush-Robin *Tarsiger indicus*	w
*	Rufous-breasted Bush-Robin *Tarsiger hyperythrus*	w 5
	Asian Magpie-Robin *Copsychus saularis*	r 1
	Hodgson's Redstart *Phoenicurus hodgsoni*	w
+	Blue-fronted Redstart *Phoenicurus frontalis*	w 1
	Plumbeous Redstart *Rhyacornis fuliginosus*	r
R	White-tailed Robin *Cinclidium leucurum*	s 5
E*	Purple Cochoa *Cochoa purpurea*	r? s 5
	Common Stonechat *Saxicola torquata*	w m 1
	Pied Bushchat *Saxicola caprata*	r
	Grey Bushchat *Saxicola ferrea*	r 2
	White-capped Redstart *Chaimarrornis leucocephalus*	r 1
+	Blue-capped Rock-Thrush *Monticola cinclorhyncha*	s
	Chestnut-bellied Rock-Thrush *Monticola rufiventris*	r?
	Blue Rock-Thrush *Monticola solitarius*	w m
	Blue Whistling Thrush *Myiophoneus caeruleus*	r 1
+	Plain-backed Mountain Thrush *Zoothera mollissima*	r 2
+	Long-tailed Mountain Thrush *Zoothera dixoni*	r 3
	Scaly Thrush *Zoothera dauma*	r? w?
R*	Long-billed Thrush *Zoothera monticola*	r? w? 5
*	Pied Ground Thrush *Zoothera wardii*	s
	Orange-headed Ground Thrush *Zoothera citrina*	r? w?
+	Tickell's Thrush *Turdus unicolor*	s
+	White-collared Blackbird *Turdus albocinctus*	r? w 2
+	Grey-winged Blackbird *Turdus boulboul*	r 2
	Chestnut Thrush *Turdus rubrocanus*	w 5
	Eye-browed Thrush *Turdus obscurus*	w 5
	Dark-throated Thrush *Turdus ruficollis*	w 1
	Little Forktail *Enicurus scouleri*	r?
+	Black-backed Forktail *Enicurus immaculatus*	r
	Slaty-backed Forktail *Enicurus schistaceus*	r
	Spotted Forktail *Enicurus maculatus*	r
+	Chestnut-headed Tesia *Tesia castaneocoronata*	r 1
	Grey-bellied Tesia *Tesia cyaniventer*	r
	Brown-flanked Bush Warbler *Cettia fortipes*	s? 4
+	Aberrant Bush Warbler *Cettia flavolivacea*	r?
I	Yellow-bellied Bush Warbler *Cettia acanthizoides*	?
	(only recorded by Stevens 1924)	
+	Grey-sided Bush Warbler *Cettia brunnifrons*	r

	Grey-breasted Prinia *Prinia hodgsoni*	r
	Striated Prinia *Prinia criniger*	r
	Hill Prinia *Prinia atrogularis*	r 3
+	Grey-capped Prinia *Prinia cinereocapilla*	r?
	Common Tailorbird *Orthotomus sutorius*	r 1
	Lesser Whitethroat *Sylvia curruca*	w 5
	Golden-spectacled Warbler *Seicercus burkii*	r 2
E+	Grey-cheeked Warbler *Seicercus poliogenys*	r 5
	Chestnut-crowned Warbler *Seicercus castaniceps*	r 2
+	Grey-hooded Warbler *Seicercus xanthoschistos*	r 1
	Yellow-bellied Warbler *Abroscopus superciliaris*	r 2
	Blyth's Crowned Warbler *Phylloscopus reguloides*	r
	Greenish Warbler *Phylloscopus trochiloides*	r s
+	Large-billed Leaf Warbler *Phylloscopus magnirostris*	s
+	Orange-barred Leaf Warbler *Phylloscopus pulcher*	r w 1
	Grey-faced Leaf Warbler *Phylloscopus maculipennis*	r w 1
	Pallas's Leaf Warbler *Phylloscopus proregulus*	r w 2
	Yellow-browed Warbler *Phylloscopus inornatus*	w m
	Dusky Warbler *Phylloscopus fuscatus*	w
	Tickell's Warbler *Phylloscopus affinis*	m
	Chiffchaff *Phylloscopus collybita*	w
	Goldcrest *Regulus regulus*	r? w
V	Large Niltava *Niltava grandis*	r 3
	Small Niltava *Niltava macgrigoriae*	r
+	Rufous-bellied Niltava *Niltava sundara*	r 2
	Blue-throated Blue Flycatcher *Cyornis rubeculoides*	s
V	Pygmy Blue Flycatcher *Muscicapella hodgsoni*	r
	Verditer Flycatcher *Muscicapa thalassina*	s
R	Ferruginous Flycatcher *Muscicapa ferruginea*	s
	Asian Sooty Flycatcher *Muscicapa sibirica*	s
	Asian Brown Flycatcher *Muscicapa latirostris*	s
V+	Sapphire Flycatcher *Ficedula sapphira*	r? 5
	Slaty-blue Flycatcher *Ficedula tricolor*	w
+	Ultramarine Flycatcher *Ficedula superciliaris*	s
R	Little Pied Flycatcher *Ficedula westermanni*	r?
	Snowy-browed Flycatcher *Ficedula hyperythra*	r
V	White-gorgetted Flycatcher *Ficedula monileger*	r?
	Orange-gorgetted Flycatcher *Ficedula strophiata*	r 1
	Red-breasted Flycatcher *Ficedula parva*	w 1
	Grey-headed Flycatcher *Culicicapa ceylonensis*	r? s 1
+	Yellow-bellied Fantail *Rhipidura hypoxantha*	r 1
	White-throated Fantail *Rhipidura albicollis*	r 2
	Rusty-cheeked Scimitar-Babbler *Pomatorhinus erythrogenys*	r 1
	White-browed Scimitar-Babbler *Pomatorhinus schisticeps*	r
	Streak-breasted Scimitar-Babbler *Pomatorhinus ruficollis*	r
V+	Slender-billed Scimitar-Babbler *Xiphirhynchus superciliaris*	r 5
+	Greater Scaly-breasted Wren-Babbler *Pnoepyga albiventer*	r?
	Lesser Scaly-breasted Wren-Babbler *Pnoepyga pusilla*	r?
E*	Tailed Wren-Babbler *Spelaeornis caudatus*	r 4
	Rufous-capped Babbler *Stachyris ruficeps*	r 2
E	Golden Babbler *Stachyris chrysaea*	r 5
	Grey-throated Babbler *Stachyris nigriceps*	r 2

V*	Brown Parrotbill *Paradoxornis unicolor*	r 4
V*	Fulvous Parrotbill *Paradoxornis fulvifrons*	r
	(only recorded by Stevens 1923)	
R	Black-throated Parrotbill *Paradoxornis nipalensis*	r 3
*	Spiny Babbler *Turdoides nipalensis*	r
+	White-throated Laughing-thrush *Garrulax albogularis*	r 1
	White-crested Laughing-thrush *Garrulax leucolophus*	r 1
+	Striated Laughing-thrush *Garrulax striatus*	r 1
+	Spotted Laughing-thrush *Garrulax ocellatus*	r 3
E+	Grey-sided Laughing-thrush *Garrulax caerulatus*	r 5
	Streaked Laughing-thrush *Garrulax lineatus*	r
E+	Blue-winged Laughing-thrush *Garrulax squamatus*	r 5
V+	Scaly Laughing-thrush *Garrulax subunicolor*	r 2
+	Black-faced Laughing-thrush *Garrulax affinis*	r 1
	Chestnut-crowned Laughing-thrush *Garrulax erythrocephalus*	r 1
E	Silver-eared Mesia *Leiothrix argentauris*	r 5
	Red-billed Leiothrix *Leiothrix lutea*	r 2
R*	Fire-tailed Myzornis *Myzornis pyrrhoura*	r 3
E	Cutia *Cutia nipalensis*	r 4
E*	Black-headed Shrike-Babbler *Pteruthius rufiventer*	r 5
	White-browed Shrike-Babbler *Pteruthius flaviscapis*	r 2
+	Green Shrike-Babbler *Pteruthius xanthochloris*	r
	Black-eared Shrike-Babbler *Pteruthius melanotis*	r 2
E+	Rusty-fronted Barwing *Actinodura egertoni*	r 1
*	Hoary Barwing *Actinodura nipalensis*	r 2
	Blue-winged Minla *Minla cyanouroptera*	r 2
	Chestnut-tailed Minla *Minla strigula*	r 1
+	Red-tailed Minla *Minla ignotincta*	r 2
V*	Golden-breasted Fulvetta *Alcippe chrysotis*	r 5
	Rufous-winged Fulvetta *Alcippe castaneceps*	r 1
+	White-browed Fulvetta *Alcippe vinipectus*	r 1
+	Nepal Fulvetta *Alcippe nipalensis*	r 5
E	Chestnut-backed Sibia *Heterophasia annectans*	r 5
+	Black-capped Sibia *Heterophasia capistrata*	r 1
+	Whiskered Yuhina *Yuhina flavicollis*	r 1
+	Stripe-throated Yuhina *Yuhina gularis*	r 1
+	Rufous-vented Yuhina *Yuhina occipitalis*	r 1
E	Black-chinned Yuhina *Yuhina nigrimenta*	r? 5
	White-bellied Yuhina *Yuhina zantholeuca*	r?
+	Black-browed Tit *Aegithalos iouschistos*	r? w
	Black-throated Tit *Aegithalos concinnus*	r 1
	Yellow-browed Tit *Sylviparus modestus*	r 1
+	Rufous-vented Black Tit *Parus rubidiventris*	r 1
	Coal Tit *Parus ater*	r?
	Great Tit *Parus major*	r 1
	Green-backed Tit *Parus monticolus*	r 1
E	Black-spotted Yellow Tit *Parus spilonotus*	r 4
E	Sultan Tit *Melanochlora sultanea*	?
	(only one record Walinder and Sandgren 1982)	
	Velvet-fronted Nuthatch *Sitta frontalis*	r 2
+	White-tailed Nuthatch *Sitta himalayensis*	r 1
	Chestnut-bellied Nuthatch *Sitta castanea*	r 1

	Wallcreeper *Tichodroma muraria*	w
*	Rusty-flanked Treecreeper *Certhia nipalensis*	r
	Common Treecreeper *Certhia familiaris*	?
	(only one record Mills and Preston 1982)	
	Mrs Gould's Sunbird *Aethopyga gouldiae*	r? 4
	Green-tailed Sunbird *Aethopyga nipalensis*	r 1
	Black-throated Sunbird *Aethopyga saturata*	r 3
	Crimson Sunbird *Aethopyga siparaja*	r 2
+	Fire-tailed Sunbird *Aethopyga ignicauda*	r? w
	Streaked Spiderhunter *Arachnothera magna*	r
R*	Yellow-bellied Flowerpecker *Dicaeum melanoxanthum*	r?
	Plain Flowerpecker *Dicaeum concolor*	r
	Buff-bellied Flowerpecker *Dicaeum ignipectus*	r 1
	Oriental White-eye *Zosterops palpebrosa*	r 1
	Maroon Oriole *Oriolus traillii*	w?
	Slender-billed Oriole *Oriolus tenuirostris*	w 5? v?
	Brown Shrike *Lanius cristatus*	w m 2
	Long-tailed Shrike *Lanius schach*	r 1
+	Grey-backed Shrike *Lanius tephronotus*	w 2
	Black Drongo *Dicrurus macrocercus*	r 1
	Ashy Drongo *Dicrurus leucophaeus*	s?
	Bronzed Drongo *Dicrurus aeneus*	r 3
	Spangled Drongo *Dicrurus hottentottus*	r
	Eurasian Jay *Garrulus glandarius*	r 4
+	Yellow-billed Blue Magpie *Urocissa flavirostris*	r 1
	Red-billed Blue Magpie *Urocissa erythrorhyncha*	r 1
	Green Magpie *Cissa chinensis*	r 2
	Grey Treepie *Dendrocitta formosae*	r 1
	House Crow *Corvus splendens*	r
	Jungle Crow *Corvus macrorhynchos*	r 1
*	Spot-winged Stare *Saroglossa spiloptera*	?
	Chestnut-tailed Starling *Sturnus malabaricus*	? 2
	Common Mynah *Acridotheres tristis*	r 1
	Jungle Mynah *Acridotheres fuscus*	r 1
	House Sparrow *Passer domesticus*	r
	Eurasian Tree Sparrow *Passer montanus*	r
	Scaly-breasted Munia *Lonchura punctulata*	r
	Yellow-breasted Greenfinch *Carduelis spinoides*	r? w?
	Plain Mountain-Finch *Leucosticte nemoricola*	w
+	Dark-breasted Rosefinch *Carpodacus nipalensis*	w
	Common Rosefinch *Carpodacus erythrinus*	w
+	Dark-rumped Rosefinch *Carpodacus edwardsii*	?
	(only recorded by Stevens 1925)	
*	Spot-winged Rosefinch *Carpodacus rhodopeplus*	w? 5
+	Gold-naped Finch *Pyrrhoplectes epauletta*	r? 5
	Brown Bullfinch *Pyrrhula nipalensis*	r 3
	Little Bunting *Emberiza pusilla*	w 2
	Crested Bunting *Melophus lathami*	r 2

PHULCHOWKI MOUNTAIN
(excluding Royal Godavari Botanical Gardens)

S indicates species breeding or probably breeding in subtropical forests

	S Crested Honey Buzzard *Pernis ptilorhyncus*	r 4
	Black Kite *Milvus migrans*	m 4
	Lammergeier *Gypaetus barbatus*	v
	Oriental White-backed Vulture *Gyps bengalensis*	m 5
	Long-billed Vulture *Gyps indicus*	m 5
	Eurasian Black Vulture *Aegypius monachus*	m 5
	S Crested Serpent Eagle *Spilornis cheela*	s 3
	Northern Goshawk *Accipiter gentilis*	m? w? 5
	S Besra *Accipiter virgatus*	r 4
	Northern Sparrowhawk *Accipiter nisus*	w 4
	Crested Goshawk *Accipiter trivirgatus*	m 5
	Shikra *Accipiter badius*	m 5
	S Black Eagle *Ictinaetus malayensis*	r 3
	Booted Eagle *Hieraaetus pennatus*	v
	S Mountain Hawk-Eagle *Spizaetus nipalensis*	r 4
	Common Kestrel *Falco tinnunculus*	m 4
	Eurasian Hobby *Falco subbuteo*	m 5
	Peregrine *Falco peregrinus*	m 5
	Black Francolin *Francolinus francolinus*	s 4
	Common Hill Partridge *Arborophila torqueola*	r 3
E	S Rufous-throated Hill Partridge *Arborophila rufogularis*	r 5
	Himalayan Monal *Lophophorus impejanus*	v
	S Kalij Pheasant *Lophura leucomelana*	r 2
	Wood Snipe *Gallinago nemoricola*	
	(collected by Hodgson (1829), but no other records)	
	Great Black-headed Gull *Larus ichthyaetus*	v
+	Speckled Woodpigeon *Columba hodgsonii*	w 3
+	Ashy Woodpigeon *Columba pulchricollis*	r 3
	S Oriental Turtle Dove *Streptopelia orientalis*	r 2
	S Spotted Dove *Streptopelia chinensis*	r 2
V	S Barred Cuckoo-Dove *Macropygia unchall*	r 5
	S Wedge-tailed Green Pigeon *Treron sphenura*	r 4
	Large Hawk-Cuckoo *Hierococcyx sparverioides*	s 3
	S Grey-bellied Plaintive Cuckoo *Cacomantis passerinus*	s 3
	S Indian Cuckoo *Cuculus micropterus*	s 3
	S Common Cuckoo *Cuculus canorus*	s 3
	Oriental Cuckoo *Cuculus saturatus*	s 3
	Lesser Cuckoo *Cuculus poliocephalus*	s?
	S Drongo-Cuckoo *Surniculus lugubris*	s 3
	S Common Koel *Eudynamys scolopacea*	s 4
	S Green-billed Malkoha *Phaenicophaeus tristis*	s 4
	Collared Scops Owl *Otus lempiji*	r 4
R	Mountain Scops Owl *Otus spilocephalus*	r 2
V	S Brown Fish Owl *Ketupa zeylonensis*	r? 5
	Collared Owlet *Glaucidium brodiei*	r 3
	S Asian Barred Owlet *Glaucidium cuculoides*	r 3
	S Jungle Nightjar *Caprimulgus indicus*	s 4

	Himalayan Swiftlet *Collocalia brevirostris*	m 3?
	White-throated Needletail *Hirundapus caudacutus*	m 4
	White-vented Needletail *Hirundapus cochinchinensis*	v
	Pacific Swift *Apus pacificus*	m 4
	Little Swift *Apus affinis*	m 3
	S Red-headed Trogon *Harpactes erythrocephalus*	
	(once recorded breeding by Proud (1955), but no other records)	
	White-breasted Kingfisher *Halcyon smyrnensis*	m 4
	S Blue-bearded Bee-eater *Nyctyornis athertoni*	s 4
	S Great Barbet *Megalaima virens*	r 2
	S Golden-throated Barbet *Megalaima franklinii*	r 3
	S Blue-throated Barbet *Megalaima asiatica*	r 4
	S Speckled Piculet *Picumnus innominatus*	r 3
	S Lesser Yellow-naped Woodpecker *Picus chlorolophus*	r 3
	S Greater Yellow-naped Woodpecker *Picus flavinucha*	r 3
	S Grey-headed Woodpecker *Picus canus*	r 2
V	S Bay Woodpecker *Blythipicus pyrrhotis*	r 4
+	Darjeeling Pied Woodpecker *Dendrocopos darjellensis*	r 3
+	Crimson-breasted Pied Woodpecker *Dendrocopos cathpharius*	r 3
	Rufous-bellied Pied Woodpecker *Dendrocopos hyperythrus*	r 3
+	Brown-fronted Pied Woodpecker *Dendrocopos auriceps*	r 3
	S Fulvous-breasted Pied Woodpecker *Dendrocopos macei*	r 2
*	S Blue-naped Pitta *Pitta nipalensis*	r? m? 5
	Barn Swallow *Hirundo rustica*	m 3
	Red-rumped Swallow *Hirundo daurica*	m 3
	Nepal House-Martin *Delichon nipalensis*	m 4
	Olive-backed Pipit *Anthus hodgsoni*	w 1
	Grey Wagtail *Motacilla cinerea*	r? m? 2
	White Wagtail *Motacilla alba*	m 2
	Large Woodshrike *Tephrodornis gularis*	v
	S Bar-winged Flycatcher-shrike *Hemipus picatus*	r s 3
	S Black-winged Cuckoo-shrike *Coracina melaschistos*	s 4
	S Large Cuckoo-shrike *Coracina novaehollandiae*	r 3
	S Scarlet Minivet *Pericrocotus flammeus*	r 2
	S Long-tailed Minivet *Pericrocotus ethologus*	r 1
E	S Grey-chinned Minivet *Pericrocotus solaris*	r? w? 4
+	Striated Bulbul *Pycnonotus striatus*	r 3
	Black-crested Bulbul *Pycnonotus melanicterus*	v
	S White-cheeked Bulbul *Pycnonotus leucogenys*	r 1
	S Red-vented Bulbul *Pycnonotus cafer*	r 1
	Mountain Bulbul *Hypsipetes mcclellandii*	r 2
	Ashy Bulbul *Hypsipetes flavalus*	m 5
	S Black Bulbul *Hypsipetes madagascariensis*	r 1
	S Common Iora *Aegithina tiphia*	r? s? 4
*	Pied Ground Thrush *Zoothera wardii*	s 4
	S Orange-bellied Leafbird *Chloropsis hardwickii*	r 3
	Northern Wren *Troglodytes troglodytes*	w 5
	Maroon-backed Accentor *Prunella immaculata*	w 5
+	Rufous-breasted Accentor *Prunella strophiata*	w 3
R	White-browed Shortwing *Brachypteryx montana*	r? m? 5
	White-tailed Rubythroat *Luscinia pectoralis*	m 5
+	Indian Blue Robin *Luscinia brunnea*	s? m? 3

	Orange-flanked Bush-Robin *Tarsiger cyanurus*	w 2
+	Golden Bush-Robin *Tarsiger chrysaeus*	w 3
+	White-browed Bush-Robin *Tarsiger indicus*	w 4
	Rufous-breasted Bush-Robin *Tarsiger hyperythrus*	v
	S Asian Magpie-Robin *Copsychus saularis*	r 3
	Blue-capped Redstart *Phoenicurus caeruleocephalus*	w 5
	Black Redstart *Phoenicurus ochruros*	w? m? 4
	Hodgson's Redstart *Phoenicurus hodgsoni*	w 3
+	Blue-fronted Redstart *Phoenicurus frontalis*	w 2
	White-throated Redstart *Phoenicurus schisticeps*	v
	Plumbeous Redstart *Rhyacornis fuliginosus*	w 2
	White-bellied Redstart *Hodgsonius phoenicuroides*	m 4
R	White-tailed Robin *Cinclidium leucurum*	s 3
	Common Stonechat *Saxicola torquata*	w m 2
	Pied Bushchat *Saxicola caprata*	r 3
	Grey Bushchat *Saxicola ferrea*	r 2
	White-capped Redstart *Chaimarrornis leucocephala*	w 2
	Blue-capped Rock-Thrush *Monticola cinclorhyncha*	m 3
	Chestnut-bellied Rock-Thrush *Monticola rufiventris*	r? w? 3
	S Blue Whistling-Thrush *Myiophoneus caeruleus*	r 3
+	Plain-backed Mountain Thrush *Zoothera mollissima*	w 4
	Scaly Thrush *Zoothera dauma*	s 3
R*	Long-billed Thrush *Zoothera monticola*	w 5
	S Orange-headed Ground Thrush *Zoothera citrina*	s 4
+	Tickell's Thrush *Turdus unicolor*	s 3
+	White-collared Blackbird *Turdus albocinctus*	w 3
+	Grey-winged Blackbird *Turdus boulboul*	r 3
	Chestnut Thrush *Turdus rubrocanus*	w 5
	Dark-throated Thrush *Turdus ruficollis*	w 1
	Little Forktail *Enicurus scouleri*	w? m? 5
	S Spotted Forktail *Enicurus maculatus*	r 3
+	Chestnut-headed Tesia *Tesia castaneocoronata*	r 1
	S Grey-bellied Tesia *Tesia cyaniventer*	r 3
+	Aberrant Bush Warbler *Cettia flavolivacea*	w 2
+	Grey-sided Bush Warbler *Cettia brunnifrons*	w 2
	S Striated Prinia *Prinia criniger*	r? s? 4
	S Common Tailorbird *Orthotomus sutorius*	r 3
	Golden-spectacled Warbler *Seicercus burkii*	w 1
	Grey-cheeked Warbler *Seicercus poliogenys*	v
	Chestnut-crowned Warbler *Seicercus castaniceps*	r 3
+	S Grey-hooded Warbler *Seicercus xanthoschistos*	r 1
	Yellow-bellied Warbler *Abroscopus superciliaris*	v
+	Black-faced Warbler *Abroscopus schisticeps*	r 3
	Blyth's Crowned Warbler *Phylloscopus reguloides*	r s 1
	Western Crowned Warbler *Phylloscopus occipitalis*	v
	Green Warbler *Phylloscopus nitidus*	m 4
	Greenish Warbler *Phylloscopus trochiloides*	m 1
	Large-billed Leaf Warbler *Phylloscopus magnirostris*	m 4
+	Orange-barred Leaf Warbler *Phylloscopus pulcher*	r 1
	Grey-faced Leaf Warbler *Phylloscopus maculipennis*	w 1
	Pallas's Leaf Warbler *Phylloscopus proregulus*	w 1
	Yellow-browed Warbler *Phylloscopus inornatus*	w 1

	Tickell's Warbler *Phylloscopus affinis*	m 3
	Goldcrest *Regulus regulus*	w 4
V	Large Niltava *Niltava grandis*	r 4
	S Small Niltava *Niltava macgrigoriae*	r 3
+	Rufous-bellied Niltava *Niltava sundara*	r 3
	Blue-throated Blue Flycatcher *Cyornis rubeculoides*	m 5
V	Pygmy Blue Flycatcher *Muscicapella hodgsoni*	r 4
	S Verditer Flycatcher *Muscicapa thalassina*	s 1
	Asian Sooty Flycatcher *Muscicapa sibirica*	s 2
	Rufous-tailed Flycatcher *Muscicapa ruficauda*	m 4
	S Asian Brown Flycatcher *Muscicapa latirostris*	s 4
	Slaty-blue Flycatcher *Ficedula tricolor*	w? m 3
+	Ultramarine Flycatcher *Ficedula superciliaris*	s 3
R	S Little Pied Flycatcher *Ficedula westermanni*	s 4
+	Slaty-backed Flycatcher *Ficedula hodgsonii*	w 5
	Snowy-browed Flycatcher *Ficedula hyperythra*	s 4
	White-gorgetted Flycatcher *Ficedula monileger*	v
	Orange-gorgetted Flycatcher *Ficedula strophiata*	w 2
	Red-breasted Flycatcher *Ficedula parva*	w 2
	S Grey-headed Flycatcher *Culicicapa ceylonensis*	r 2
+	Yellow-bellied Fantail *Rhipidura hypoxantha*	w 3
	S White-throated Fantail *Rhipidura albicollis*	r 2
	S Rusty-cheeked Scimitar-Babbler *Pomatorhinus erythrogenys*	r 2
	S Streak-breasted Scimitar-Babbler *Pomatorhinus ruficollis*	r 3
+	Greater Scaly-breasted Wren-Babbler *Pnoepyga albiventer*	w 3
	Lesser Scaly-breasted Wren-Babbler *Pnoepyga pusilla*	r 3
+	S Black-chinned Babbler *Stachyris pyrrhops*	r 2
	S Grey-throated Babbler *Stachyris nigriceps*	r 3
R	Black-throated Parrotbill *Paradoxornis nipalensis*	r 4
*	S Spiny Babbler *Turdoides nipalensis*	r 3
+	White-throated Laughing-thrush *Garrulax albogularis*	r 1
	S White-crested Laughing-thrush *Garrulax leucolophus*	r 1
+	S Striated Laughing-thrush *Garrulax striatus*	r 2
V+	S Rufous-chinned Laughing-thrush *Garrulax rufogularis*	r 3
E+	Grey-sided Laughing-thrush *Garrulax caerulatus*	r 3
	Streaked Laughing-thrush *Garrulax lineatus*	r 4
E+	S Blue-winged Laughing-thrush *Garrulax squamatus*	r 5
	Chestnut-crowned Laughing-thrush *Garrulax erythrocephalus*	r 1
	S Red-billed Leiothrix *Leiothrix lutea*	r 3
E	Cutia *Cutia nipalensis*	r 4
	White-browed Shrike-Babbler *Pteruthius flaviscapis*	r 3
+	Green Shrike-Babbler *Pteruthius xanthochloris*	r 4
	Black-eared Shrike-Babbler *Pteruthius melanotis*	r 3
*	Hoary Barwing *Actinodura nipalensis*	r 3
	S Blue-winged Minla *Minla cyanouroptera*	r 3
	Chestnut-tailed Minla *Minla strigula*	r 2
+	Red-tailed Minla *Minla ignotincta*	r 5
	Rufous-winged Fulvetta *Alcippe castaneceps*	r 2
+	White-browed Fulvetta *Alcippe vinipectus*	r 1
+	S Nepal Fulvetta *Alcippe nipalensis*	r 2
+	Black-capped Sibia *Heterophasia capistrata*	r 1
+	S Whiskered Yuhina *Yuhina flavicollis*	r 1

+	Stripe-throated Yuhina *Yuhina gularis*	r 1
+	Rufous-vented Yuhina *Yuhina occipitalis*	w 4
	S White-bellied Yuhina *Yuhina zantholeuca*	r 2
	S Black-throated Tit *Aegithalos concinnus*	r 1
	Yellow-browed Tit *Sylviparus modestus*	r b 3
	S Great Tit *Parus major*	r 4
	Green-backed Tit *Parus monticolus*	r 2
	S Black-lored Tit *Parus xanthogenys*	r 2
	S Velvet-fronted Nuthatch *Sitta frontalis*	r 3
+	White-tailed Nuthatch *Sitta himalayensis*	r 1
	S Chestnut-bellied Nuthatch *Sitta castanea*	r 1
	Wallcreeper *Tichodroma muraria*	w 4
	Brown-throated Treecreeper *Certhia discolor*	r 2
*	Rusty-flanked Treecreeper *Certhia nipalensis*	r? w? 4
	Fire-capped Tit *Cephalopyrus flammiceps*	v
	Mrs Gould's Sunbird *Aethopyga gouldiae*	w 4
	Green-tailed Sunbird *Aethopyga nipalensis*	r 1
	S Black-throated Sunbird *Aethopyga saturata*	r 3
+	Fire-tailed Sunbird *Aethopyga ignicauda*	w 2
R*	Yellow-bellied Flowerpecker *Dicaeum melanoxanthum*	w 3
	S Plain Flowerpecker *Dicaeum concolor*	r s 3
	S Buff-bellied Flowerpecker *Dicaeum ignipectus*	r 2
	S Oriental White-eye *Zosterops palpebrosa*	r 2
	S Maroon Oriole *Oriolus traillii*	r 3
	S Long-tailed Shrike *Lanius schach*	r 2
+	Grey-backed Shrike *Lanius tephronotus*	w 3
	S Black Drongo *Dicrurus macrocercus*	r 1
	S Ashy Drongo *Dicrurus leucophaeus*	r 2
	S Bronzed Drongo *Dicrurus aeneus*	r s 3
	S Lesser Racket-tailed Drongo *Dicrurus remifer*	s 4
	Eurasian Jay *Garrulus glandarius*	r 1
+	Lanceolated Jay *Garrulus lanceolatus*	r 4
+	Yellow-billed Blue Magpie *Urocissa flavirostris*	r 5
	S Red-billed Blue Magpie *Urocissa erythrorhyncha*	r 2
	Rufous Treepie *Dendrocitta vagabunda*	v
	(probably escaped birds)	
	S Grey Treepie *Dendrocitta formosae*	r 2
	House Crow *Corvus splendens*	m 4
	S Jungle Crow *Corvus macrorhynchos*	r 3
	S Common Mynah *Acridotheres tristis*	r 2
	S Jungle Mynah *Acridotheres fuscus*	r 2
	Eurasian Tree Sparrow *Passer montanus*	r 2
	S Scaly-breasted Munia Lonchura *punctulata*	r 3
	Brambling *Fringilla montifringilla*	v
	Tibetan Serin *Serinus thibetanus*	w 3
	Red-browed Finch *Callacanthis burtoni*	v
	Yellow-breasted Greenfinch *Carduelis spinoides*	w 2
+	Dark-breasted Rosefinch *Carpodacus nipalensis*	w 2
	Common Rosefinch *Carpodacus erythrinus*	w 2
+	Pink-browed Rosefinch *Carpodacus rhodochrous*	w 3
*	Crimson-browed Finch *Propyrrhula subhimachala*	w 5
V*	Scarlet Finch *Haematospiza sipahi*	w 4

+	Gold-naped Finch *Pyrrhoplectes epauletta*	w 4
	Brown Bullfinch *Pyrrhula nipalensis*	r 3
*	Spot-winged Grosbeak *Mycerobas melanozanthos*	w 3
	Little Bunting *Emberiza pusilla*	w 3

REFERENCES

Ali, S. and Ripley, S. D. (1984) *Handbook to the birds of India and Pakistan.*
Compact edition. Bombay: Oxford University Press.

Alind, P. (1986) Notes on birds recorded in Nepal, February-March 1986.
Unpublished.

Andrews, T. (1986) Notes on birds recorded in northern India and Nepal, winter
1984-85. Unpublished.

Anon. (1987) On not planting trees. *Banko Janakari* 1(1): 2.

Axel-Bauer, C. (1986) Birds and mammals seen on 'Skof' - tour to India and
Nepal 19 April - 9 May 1986. Unpublished.

Baker, T. (1981) Notes on birds recorded in Nepal, 1981. Unpublished.

Baker, T. (1983) Notes on birds recorded in Nepal, February-March 1983.
Unpublished.

Beaman, M. A. S. (1982) Notes on birds recorded in Nepal, 1973-82. Unpublished.

Berg, A. B. van den and Bosman, C. A. W. (1976) List of birds observed in
Nepal, April 1976. Unpublished.

Biswas, B. (1974) Zoological results of the Daily Mail Himalayan Expedition
1954: notes on some birds of Eastern Nepal. *J. Bombay nat. Hist. Soc.* 71:
456-495.

Blanchon, J.-J. and Dubois, P. (1987) Voyage au Nepal, Mars 1987.
Unpublished.

Bolding, J. and Jorgensen, T. (1987) List of birds recorded in Nepal and India,
October 1986-January 1987. Unpublished.

Bolton, M. (1976) Lake Rara National Park Management Plan 1976-81.
FO/NEP/72/002 Project Working Document No. 3. Kathmandu: UNDP/FAO.

Borradaile, L. J., Green, M. J. B., Moon, L. C., Robinson, P. J. and Tait, A.
(1977) Langtang National Park Management Plan 1977-82 FO NEP/72/002
Field Document No. 7. Kathmandu: DUHE/HMG/UNDP.FAO.

Bradbear, P. (1986) Notes on birds recorded in Nepal, September-October 1986.
Unpublished.

Byrne, R. W. and Harris, S. M. (1975) Skeletal report on birds and mammals
seen during September-November 1975 in Nepal. Unpublished.

Carson, B. (1985) *Erosion and sedimentation processes in the Nepalese Himalaya.*
ICIMOD Occasional Paper No. 1 Kathmandu: International Centre for
Integrated Mountain Development.

Carter, J. (1987) Organisations concerned with forestry in Nepal. FRIC
Occasional Paper No. 2/87. Kathmandu: Forestry Research and Information
Centre.

Christensen, S., Bijlsma, R., Roder de, F. and Henriksen, M. (1984) Notes on
birds recorded in Nepal, 1984. Unpublished.

Clements, A. and Bradbear, N. (1981) Systematic list of species seen in Nepal
and India, November-December 1981. Unpublished.

Clugston, D. L. (1985) A checklist of the birds and mammals seen in Nepal during the twenty-three day period from 8-30 March 1985. Unpublished.

Cocker, M. and Inskipp, C. (1988) *A Himalayan ornithologist. The life and work of Brian Houghton Hodgson.* Oxford: Oxford University Press.

Collar, N. J. and Stuart, S. N. (1985) *Threatened birds of Africa and related islands.* The ICBP/IUCN Red Data Book, Part 1, Third edition. Cambridge: ICBP/IUCN.

Corbett, G. B. (1974) Birds recorded on the RAF Dhaulagiri Expedition, March-- May 1975. Unpublished.

Couronne, B. and Kovacs, J. C. (1986). Observations ornithologiques au Nepal, Février-Mars 1986. Unpublished.

Cox, J. Jr (1978) Avian Jottings for Nepal, 1978. Unpublished.

Cox, J. Jr (1982) Avian checklist of species observed during 1978 in the district of Kapilvastu, central tarai. Unpublished.

Cox, J. Jr (1984) Notes on birds recorded in Nepal, 1984. Unpublished.

Cox, J. Jr (1985) Selected notes of a brief avian survey of Royal Bardia Wildlife Reserve and periphery, west Nepal during November 1985. Unpublished.

Cox, J. Jr (1985) Checklist of birds recorded within Rara Lake National Park, May 1985. Unpublished.

Cox, J. Jr (1985) Further notes on birds recorded in Nepal, 1976-84. Unpublished.

Cox, J. Jr (1985) Birds of the Rara-Jumla area, west Nepal, May 1985. Unpublished.

Cronin, E. W. Jr (1979) *The Arun, a natural history of the world's deepest valley.* Boston: Houghton Mifflin.

Cronin, E. W. Jr and Sherman, P. W. (1976) A resource-based mating system: The Orange-rumped Honeyguide. *The Living Bird* 15: 5-32.

Dahmer, T. A. (1976) Birds of Kosi Tappu Wildlife Reserve. Unpublished.

Dawson, I. (1983) Notes on birds recorded in Nepal, December 1982-January 1983. Unpublished.

del-Nevo, A. and Ewins, P. (1981) Bird Watching in Nepal, 7th December 1980-19th February 1981. Unpublished.

Diesselhorst, G. (1968) Beiträge zur ökologie der Vögel Zentral - und Ost - Nepals. *Khumbu Himal* 2: 1-417.

Dinerstein, E. (1979) An ecological survey of the Royal Karnali-Bardia Wildlife Reserve, Nepal. Unpublished thesis.

Dixit, K. (1986) Hill of flowers. Pp. 85-90 in K. Dixit and L. Tuting (eds.) *Bikas-binas/development-destruction?* Munich: Geobuch.

Dobremez, J. F. (1976) Le Népal écologie et biogéographie. Paris: Centre National de la Recherche Scientifique.

Dobremez, J. F. (1984) *Carte écologique du Népal. Region Dhangarhi-Api 1/250,000.* Documents de Cartographie écologique, Grenoble.

Dobremez, J. F., Jest, C., Toffin, G., Vartanian, M. C. and Vigny, F. (1974) *Carte écologique du Népal. Region Kathmandu-Everest 1/250,000.* Documents de Cartographie écologique, Grenoble.

Dobremez, J. F., Joshi, D. P., Bottner, P., Jest, C. and Vigny, F.(1984) *Carte écologique du Népal. Region Butwal-Mustang 1/250,000.* Documents de Cartographie écologique, Grenoble.

Dobremez, J. F., Joshi, D. P., Shrestha, T. B. and Vigny, F. (1985) *Carte écologique du Népal. Region Nepalganj-Dailekh 1/250,000.* Documents de Cartographie écologique, Grenoble.

Dobremez, J. F. and Shakya, P. R. (1977) *Carte écologique du Népal. Region Biratnagar-Kangchenjunga 1/250,000.* Documents de Cartographie écologique, Grenoble.

Dobremez, J. F., Shrestha, B. K., Verniau, S. and Vigny, F. (1974) *Carte écologique du Népal. Region Terai central 1/250,000.* Documents de Cartographie écologique, Grenoble.

Dobremez, J. F. and Shrestha, T. B. (1980) *Carte écologique du Népal. Region Jumla-Saipal 1/250,000.* Documents de Cartographie écologique, Grenoble.

Dymond, J. N. (1986) Selected bird list, Nepal, February-March 1986. Unpublished.

Eames, J. (1982) Notes on birds recorded in Nepal, 1982. Unpublished.

Eames, J. and Grimmett, R. F. (1982) Birds recorded at Kosi Tappu Wildlife Reserve, April 1982. Unpublished.

Eve, V. and Hibberd, G. (1987) Notes on birds recorded in Nepal, 1987. Unpublished.

Fairbank, R. J. (1980) Notes on birds recorded in Nepal, November 1979-January 1980. Unpublished.

Fairbank, R. J. (1982) Notes on birds recorded in Nepal, 1982. Unpublished.

Farrow, D. (1982) Notes on birds recorded in Nepal, March-April 1982. Unpublished.

Fleming, R. L. Jr (1968) The Waxwing, *Bombycilla garrulus* (Linnaeus) in Nepal. *J. Bombay nat. Hist. Soc.* 65: 488.

Fleming, R. L. Jr (1979) Notes on birds seen in the Everest National Park, Nepal, May 1979. Unpublished.

Fleming, R. L. Jr (1981) Distribution information on various bird species in Nepal. Pers. comm. March 1981.

Fleming, R. L. Jr. (1982) List of birds recorded in Dolpo district in 1971. Unpublished.

Fleming, R. L. Sr and Fleming, R. L. Jr (1970) *Birds of Kathmandu Valley and surrounding hills: a check list.* Kathmandu.

Fleming, R. L. Sr, Fleming, R. L. Jr and Bangdel, L. S. (1984) *Birds of Nepal.* Third edition. Kathmandu: Avalok.

Fleming, R. L. Sr and Traylor, M. A. (1961) Notes on Nepal birds. *Fieldiana: zool.* 35(9): 447-487.

Fleming, R. L. Sr and Traylor, M. A. (1964) Further notes on Nepal birds. *Fieldiana: zool.* 35(9): 495-558.

Fleming, R. L. Sr and Traylor, M. A. (1968) Distributional notes on Nepal birds. *Fieldiana: zool.* 53(3) 147-203.

Forster, E. (1982) *Himalayan Solo.* Shrewsbury: Anthony Nelson.

Gantlett, S. J. M. (1981) Notes on birds recorded in Nepal, November 1981. Unpublished.

Garnatt, K. J. (1981) Sagarmatha National Park Management Plan. Unpublished. Kathmandu: Department of National Parks and Wildlife Conservation.

Gawn, S. C. *in litt.* (1987).

Gawn, S. C. (1987) Birding in India and Nepal a trip report. Unpublished.

Ghimre, G. P. S. (1984-1985) Plant ecology of central midland Nepal (a case study of Godavari-Phulchowki forest). Pp. 427-437 in T. C. Majupuria (ed.) *Nepal, nature's paradise.* Bangkok: White Lotus.

Goodwin, A. (1986) Notes on birds recorded in Nepal, 1986. Unpublished.

Gould, J. (1837-1838) *Icones Avium, or figures and descriptions of new and interesting species of birds from various parts of the globe.* London.

Gray, J. E. (1863) *Catalogue of the specimens and drawings of mammals, birds, reptiles and fishes of Nepal and Tibet, presented by B. H. Hodgson, Esq. to the British Museum.* Second edition. London.

Gray, J. E. and Gray, G. R. (1846) *Catalogue of the specimens and drawings of mammalia and birds of Nepal and Thibet, presented by B. H. Hodgon, Esq. to the British Museum.* London.

Green, M. J. B. (1980) A report on conservation and management issues within the Langtang National Park. Unpublished.

Green, M. J. B. (1986) Directory of Indomalayan Protected Areas: Nepal. Draft. Cambridge: IUCN.

Grimmett, R. F. (1982) Notes on birds recorded in Nepal, 1982. Unpublished.

Gurung, K. K. (1983) *Heart of the Jungle.* London: André Deutsch.

Halberg, K. (1987) Notes on birds seen in Nepal, 1985. Unpublished.

Halberg, K. and Petersen, I. (1984) Himalaya 1978-83. Observations of birds, mammals and some reptiles. Unpublished.

Hall, J. (1981) Notes on birds recorded in Nepal, November 1980-March 1981. Unpublished.

Hall, P. (1978) Notes on birds recorded in Nepal, September 1976-December 1978. Unpublished.

Halliday, J. (1986) Notes on birds recorded in Nepal, October-December 1986. Unpublished.

Hamon, P. (1981) Bird observations in Nepal, December 1980 - January 1981. Unpublished.

Harrap, S. (1985) Birds seen on Phulchowki, Nepal in 1985. Unpublished.

Harrop, A. H. J. (1986) Notes on birds recorded in Nepal, March-April 1986. Unpublished.

Harris, E. (1978) Birds identified from Lamosangu to Everest Base Camp, October-November 1978. Unpublished.

Heath, P. J. (1986) Notes on birds recorded in Nepal, 1986. Unpublished.

Heath, P. J. and Thorns, D. M. (in press) Bristled Grass Warbler *Chaetornis striatus*: a new species and presumed first breeding record from Nepal, with some comments on identification of Bristled Grass Warbler and Large Grass Warbler *Graminicola bengalensis. Forktail* IV.

Heegaard, M., Priemé, A. and Turin, R. (1987) Northern part of the Indian subcontinent. Unpublished report.

Heinen, J. (1988) Notes on birds recorded at Kosi Tappu Wildlife Reserve, January 1987-March 1988. Unpublished.

Hines, P. (1987) Notes on birds recorded in Nepal in spring 1985. Unpublished.

HMG/IUCN (1983) National conservation strategy for Nepal: a prospectus. Gland: IUCN.

Hodgson, B. H. (1829) Notes and original watercolour paintings of the birds of Nepal, Tibet and India, held in the Zoological Society of London Library. Unpublished.

Hodgson, B. H. (1837) Indian quails. *Bengal Sporting Mag.* 9: 343-346.

Holmstrom, G. (1983) Notes on birds recorded in Nepal during 1981 and 1983. Unpublished.

Innes, R. and Lewis, P. (1984) Notes on birds recorded in Nepal, March to May 1984. Unpublished.

Inskipp, C. (1987) Trekking and the birdwatcher in Nepal. *Bull. Oriental Bird Club* 6: 16-18.

Inskipp, C. (1988) Khaptad National Park: An account of current knowledge and conservation value. Unpublished report to Dept of National Parks and Wildlife Conservation, Nepal.

Inskipp, C. and Inskipp, T. P. (1982) Notes on birds recorded in Nepal, 1982. Unpublished.

Inskipp, C. and Inskipp, T. (1983) *Report on a survey of Bengal Floricans Houbaropsis bengalensis in Nepal and India, 1982.* International Council for Bird Preservation, Study Report No. 2.

Inskipp, C. and Inskipp, T. (1985) *A guide to the birds of Nepal.* Beckenham: Croom Helm.

Inskipp, C. and Inskipp, T. P. (1986) Some important birds and forests in Nepal. *Forktail* 1: 53-64.

Inskipp, C. and Inskipp, T. P. (1986) Notes on birds recorded in Nepal, 1986. Unpublished.

Inskipp, T. P. and Inskipp, C. (1977) Notes on birds recorded in Nepal, December 1977. Unpublished.

Inskipp, T. P. and Inskipp, C. (1980) Notes on birds recorded in Nepal, April-- May 1980. Unpublished.

Inskipp, T. P. and Inskipp, C. (1981) Notes on birds recorded in Nepal, February-March 1981. Unpublished.

Inskipp, T. P. *et al.* (1971) Notes on birds recorded in Nepal, September 1970-March 1971. Unpublished.

Jackson, J. K. (1987) *Manual of afforestation in Nepal.* Kathmandu: Nepal UK Forest Research Project.

Jackson, R. (1978) A report on wildlife and hunting in the Namlang (Langu) valley of west Nepal. Report to National Parks and Wildlife Conservation Dept. Kathmandu. Unpublished.

Jepson, P. (1985) Systematic list of birds seen in Nepal, March-May 1985. Unpublished.

Johns, R. J. (1982) Notes on birds recorded in Nepal, December 1981-January 1982. Unpublished.

Jongeling, B. (1983) Notes on birds recorded in Nepal, 1983. Unpublished.

Joshi, A. R. (1986) Shivapuri Watershed and Wildlife Reserve. Unpublished.

Juliusburger, R. (1987) Notes on birds recorded in Nepal, spring 1987. Unpublished.

Justice, S. (1978) Notes on birds recorded in Nepal, 1976-78. Unpublished.

Lelliott, A. D. (1981) Notes on the birds recorded in Nepal, 1978-81. Unpublished.

Kennerley, P. (1982) Notes on birds recorded in Nepal, 1982. Unpublished.

Kenting (1986) Land resources mapping project. Kathmandu: Kenting Earth Sciences Ltd.

Khadka, R. B., Shrestha, J. and Tamrakar, A. S. (1984-1985) Ecology of Godawari hills: a case study. Pp. 408-426 in T. C. Majupuria (ed.) *Nepal, nature's paradise.* Bangkok: White Lotus.

King, W. B. (1978-1979) *Red data book, 2: Aves.* 2nd edition. Morges, Switzerland: IUCN.

King, B. (1980) Notes on birds recorded in Nepal 1972-79. Unpublished.

Kjellen, N., Jirle, E. and Walinder, G. (1981) Asien-81. Unpublished.

Klapste, J. (1986) List of birds seen on Langtang trek, north-central Nepal, March-April 1986. Unpublished.

Kovacs, J. C. (1987) Compte rendu 'un voyage naturaliste au Nepal. Février-- Mars 1987. Unpublished.

Krabbe, E. (1983) List of bird specimens in the Zoological Museum of Copenhagen, collected by G. B. Gurung, S. Rana and P. W. Soman from Nepal, 1959. Unpublished.

Lambert, F. (1979) Notes on birds recorded in Nepal, 1978-79. Unpublished.

Lelliott, A. D. and Yonzon, P. B. (1979) Pheasant studies in Annapurna Himal (1) and (2). *Proc. 1st Intern. Pheasant Symposium.* Kathmandu 1979: 56-62.

Lister, V. (1979) Notes on birds recorded in Nepal, April-June 1979. Unpublished.

Madge, S. C. (1986) Selected notes on Nepal Birdquest tour, February-March 1986. Unpublished.

Madge, S. C. and Madge, P. (1982) Notes on birds recorded in Nepal, April-May 1982. Unpublished.

Mahat, T. B. S., Griffin, D. M. and Shepherd, K. R. (1986) Human impact on some forests of the middle hills of Nepal 1. Forestry in the context of the traditional resources of the state. *Mountain Research and Development* 6(3): 223-232.

Martens, J. (1971) Zur Kenntnis des Vogelzuges im nepalischen Himalaya. *Vogelwarte* 26: 113-128.

Martens, J. (1972) Brutverbreitung paläarktischer Vögel im Nepal-Himalaya. *Bonn. zool. Beitr.* 23: 95-121.

Martens, J. (1975) Akustische Differenzierung verwandtschaftlicher Beziehungen in der Parus Gruppe nach Untersuchungen im Nepal-Himalaya. *J. Orn.* 116(4): 369-433.

Martens, J. (1979) Die Fauna des Nepal-Himalaya - Entstehung und Erforschung. *Natur und Museum* 109(7): 221-243.

Martens, J. (1983) Forests and their destruction in the Himalayas of Nepal. Misc. Papers, Nepal Res. Centre, 35: 1-70; Kathmandu and Wiesbaden.

Martins, R. P. (1982) Birds seen in Khumbu National Park from Lukla northwards, May 1982. Unpublished.

Mayer, S. (1986) Notes on birds recorded in Nepal, October 1985-April1986. Unpublished.

Melville, D. S. and Hamilton, V. J. (1981) Notes on birds recorded in Nepal, November-December 1981. Unpublished.

Mills, D. G. H. (1987) Notes on birds recorded in Nepal, 1981-87. Unpublished.

Mills, D. G. H. and Preston, N. A. (1981) Notes on birds recorded in Nepal, 1981. Unpublished.

Mills, D. G. H., Preston, N. A. and Winyard, C. (1982) Notes on birds recorded in Nepal. Unpublished.

Mishra, H. (1986) Deforestation and environmental degradation through tourism in Nepal. Pp. 320-323 in K. Dixit and L. Tuting (eds.). Bikas-binas/development-destruction? Munich: Geobuch.

Morioka, H. (1985) Notes on birds of Dhorpatan, central Nepal. *Tori* 33(4): 113-122.

Munthe, K. (1981) Notes on birds recorded in Nepal, December 1980-January 1981. Unpublished.

Murphy, C. (1986) Notes on birds recorded on Sheopuri, 1986. Unpublished.

NAFP (1979) *Nepal's national forestry plan, 1976 (2033).* An unofficial English translation. Kathmandu: Nepal Australia Forestry Project.

Nepali, H. S. (1986) List of Nepalese bird specimens. Unpublished.

Nepali, H. S. (1986) Notes on birds recorded in the Barun valley. Unpublished.

Nickel, H. and Trost, R. (1983) Vogelkundliche beobachtungen einer reise nach Indien und Nepal, January-April 1983. Unpublished.

Nicolle, S. (1987) Notes on birds recorded in Nepal, 1987. Unpublished.

Nilsson, T. (1982) Notes on birds recorded in Nepal, 1982. Unpublished.

Oy, J. P. and Madecor (1987a) Master plan for forestry sector Nepal: National Parks and Wildlife Development Plan. Unpublished draft prepared for Ministry of Forestry and Soil Conservation.

Oy, J. P. and Madecor (1987b) Master plan for forestry sector Nepal: The Forest Resources of Nepal. Unpublished draft prepared for Ministry of Forestry and Soil Conservation.

Parr, M. (1982) Notes on birds recorded in Nepal, 1982. Unpublished.

Petersen, I. (1983) Notes on birds recorded in Nepal, 1980. Unpublished.

Polunin, O. (1952) Notes on birds recorded in Nepal, 1951-52. Unpublished.

Polunin, O. (1955) Some birds collected in Langtang Khola, Rasua Garhi district, central Nepal. *J. Bombay nat. Hist. Soc.* 52: 856-896.

Porter, R. F., Oddie, W. E. and Marr, B. A. E. (1981) Notes on birds recorded in Nepal, February 1981. Unpublished.

Powell, N. and Pierce, R. (1984) Notes on birds recorded in Nepal, 14 March to 30 April 1984. Unpublished.

Pritchard, D. E. (1980) The birds of western Nepal: the report of the ornithologists. In Saipal 79. Univ. of Durham Expedition to western Nepal 1979. Unpublished.

Pritchard, D. E. and Brearey, D. (1985) Saipal 82-83, Biological survey of Nepal's far western hills. 3. Birds and other wildlife of Lake Rara National Park, north-west Nepal. Unpublished.

Proud, D. (1949) Some notes on the birds of the Nepal Valley. *J. Bombay nat. Hist. Soc.* 48: 695-719.

Proud, D. (1952) Some birds seen on the Gandak-Kosi watershed in March 1951. *J. Bombay nat. Hist. Soc.* 50: 355-365.

Proud, D. (1952) Further notes on the birds from Nepal Valley. *J. Bombay nat. Hist. Soc.* 50: 667-670.

Proud, D. (1955) More notes on the birds from Nepal Valley. *J. Bombay nat. Hist. Soc.* 53: 57-78.

Proud, D. (1959) Notes on the Spiny Babbler, *Acanthoptila nipalensis* (Hodgson), in the Nepal Valley. *J. Bombay nat. Hist. Soc.* 56: 330-332.

Proud, D. (1961) Notes on the birds from Nepal. *J. Bombay nat. Hist. Soc.* 58: 798-805.

Rana, P. S. J. B., Pandey, N. R. and Mishra, H. P. (1984) An introduction to the King Mahendra Trust for Nature Conservation. King Mahendra Trust for Nature Conservation Series No. 1/84.

Rand, A. L. and Fleming, R. L. (1957) Birds of Nepal. *Fieldiana: zool.* 41: 1-218.

Redman, N. J. and Murphy, C. (1979) Notes on birds recorded in Nepal, December 1978 - June 1979. Unpublished.

Reid, T. (1984) Notes on birds recorded in Nepal, winter 1983-84. Unpublished.

Rice, C. (1978) Notes on birds seen on a trek to Muktinath, April 1978. Unpublished.

Richards, G. and Richards, L. (1981) Notes on birds recorded in Nepal, February-April 1981. Unpublished.

Rieger, H. C. (1981) Man versus mountain. Pp. 351-376 in D. D. Bhatt (ed.) *The Himalaya, aspects of change.* Delhi: Oxford University Press.

Riessen, A. van (1986) Birds of far western Nepal, 1983-85. Unpublished.

Roberts, J. O. M. (1978) Breeding of the Mallard (*Anser platyrhynchos*) in Nepal. *J. Bombay nat. Hist. Soc.* 75: 485-486.

Roberts, J. O. M. (1979) Report on Status of Pheasants: Nepal. *Proc. 1st Intern. Pheasant Symposium* Kathmandu 1979: 22-26.

Roberts, T. J. and King, B. (1986) Vocalisations of the owl genus *Otus* in Pakistan. *Ornis Scandinavica* 17: 299-305.

Robinson, P. (1977) Notes on birds recorded in Nepal, May 1977. Unpublished.

Robinson, T. (1986) Records of Kessler's Thrush *Turdus kessleri* in Nepal, January-February 1986. Unpublished.

Robson, C. (1979) Notes on birds recorded in Nepal, November-December 1979. Unpublished.

Robson, C. (1982) Notes on birds recorded in Nepal, 1982. Unpublished.

Roder de, F. (in press) The migration of raptors, south of Annapurna, Nepal, autumn 1985. *Forktail* IV.

Rooke, S. (1982) Notes on birds recorded in Nepal, December 1982. Unpublished.

Ross, J. (1983) Notes on birds recorded in Nepal, January-April 1983. Unpublished.

Rossetti, J. B. O. (1978) Notes on birds recorded in Nepal, August 1978. Unpublished.

Russell, V. (1981) Notes on birds recorded in Nepal, 1981. Unpublished.

Sakya, K. (1978) *Dolpo, The world behind the Himalayas.* Kathmandu: Sharda Prakashan Grika.

Sattaur, O. (1987) Trees for the people. *New Scientist* 10 September 1987: 58-62.

Schaaf, D., Rice, C. and Fleming, R. L. Jr. (1977) A partial checklist of the birds of Sukla Phanta Wildlife Reserve, Nepal. Unpublished.

Scharringa, J. (1987) Ornithological observations Nepal, December 1986-January 1987. Unpublished.

Searle, M. (1980) Notes on birds recorded in Nepal, 1980. Unpublished.

Shakya, R. and Thompson, I. (1987) A greater role for broadleaved species in the middle hills. *Banko Janakari* 1(2): 18-20.

Sharpe, R. B. (1883) *Catalogue of the birds in the collection of the British Museum,* Volume 7 - Timaliidae (part).

Shrestha, T. B. (1984-1985) Vegetation and people of western Nepal with special reference to Karnali zone. Pp. 360-368 in T. C. Majupuria (ed.) *Nepal, nature's paradise.* Bangkok: White Lotus.

Shrestha, T. B., Dhungel, S. and Davis, R. (1985) The Barun Valley Report. Unpublished.

Sieurin, P. (1987) Record of Bean Goose *Anser fabalis* in Nepal, December 1985. Unpublished.

Simpson, N. (1985) Notes on birds recorded in Nepal, 1985. Unpublished.

Singha, I. L. (1984-1985) Rainfall distribution. Pp. 56-58 in T. C. Majupuria (ed.) *Nepal, nature's paradise.* Bangkok: White Lotus.

Smith, S. (1988) Notes on birds recorded at Royal Bardia Wildlife Reserve, February 1988. Unpublished.

Smythies, B. E. (1947) Some birds of the Gandak-Kosi watershed including the pilgrim trail to the sacred lake of Gosainkund. *J. Bombay nat. Hist. Soc.* 47: 432-443.

Smythies, B. E. (1950) More notes on the birds of the Nepal Valley. *J. Bombay nat Hist. Soc.* 49: 513-518.

Stainton, J. D. A. (1972) *Forests of Nepal.* London: John Murray.

Stevens, H. (1923) Notes on the birds of the Sikkim Himalayas, Part 2. *J. Bombay nat. Hist. Soc.* 29: 723-740.

Stevens, H. (1924a)   Notes on the birds of the Sikkim Himalayas, Part 3.   *J. Bombay nat. Hist. Soc.* 29: 1007-1030.

Stevens, H. (1924b)   Notes on the birds of the Sikkim Himalayas, Part 4.   *J. Bombay nat. Hist. Soc.* 30: 54-71.

Stevens, H. (1925a)   Notes on the birds of the Sikkim Himalayas, Part 5.   *J. Bombay nat. Hist. Soc.* 30: 352-379.

Stevens, H. (1925b)   Notes on the birds of the Sikkim Himalayas, Part 6.   *J. Bombay nat. Hist. Soc.* 30: 664-685.

Stevens, H. (1925c)   Notes on the birds of the Sikkim Himalayas, Part 7.   *J. Bombay nat. Hist. Soc.* 30: 872-893.

Suter, W. (1983)   Ornithological and mammalogical observations in Nepal and NW India, including a few observations in Bangladesh and Pakistan, December 1982-February 1983. Unpublished.

Thiollay, J. M. (1977)   Notes on birds reocrded in Nepal. Unpublished.

Thiollay, J. M. (1978)   Distributions des Falconiformes nicheurs autour du massif de l'Annapurna (Himalaya Central). *L'Oiseau et R.F.O.* 48:291-310.

Thiollay, J. M. (1978)   Ornithological survey of Royal Chitwan National Park, October-November 1978. Unpublished.

Thiollay, J. M. (1980)   L'evolution des peuplements d'oiseaux le long d'ungradient altitudinal dans l'Himalaya Central. *Rev. Ecol. (Terre et Vie)* 34: 199-269.

Thorns, D. (1987)   Notes on birds recorded in Nepal, 1986. Unpublished.

Turton, J. M. and Speight, G. J. (1982)   A report on birds seen in Nepal, 1982. Unpublished.

Underwood, L. (1979)   Notes on birds seen at Sheopuri, Kathmandu Valley, Nepal, March 13 1979. Unpublished.

Underwood, L. (1979)   North of Pokhara during the monsoon (Part I). *Nepal Nature Conservation Society Newsletter* September 1979.

Underwood, L. (1979)   North of Pokhara during the monsoon (Part II). *Nepal Nature Conservation Society Newsletter* October 1979.

Walinder, G. and Sandgren, B. (1983)   Artlista over faglar observerade i Nepal, 10.3-12.4 1982. Unpublished.

Warwick, J. (1985)   Selected bird records for Nepal, April-May 1985. Unpublished.

West Nepal Adventures (P) Ltd (1988)   Notes on birds recorded at Royal Bardhia Wildlife Reserve, 1988. Unpublished.

Wolstencroft, J. A. (1981)   Notes on birds recorded in Nepal, March-May 1981. Unpublished.

Wolstencroft, J. A. (1982)   Notes on birds recorded in Nepal, 1982. Unpublished.

World Bank (1978)   Nepal forestry sector review.   Report no. 1962-NEP. Washington: World Bank.

Wotham, M. and Bond, G. (1984)   Notes on birds recorded in Nepal, 1984. Unpublished.

LIST OF FOREST TYPES AND THEIR BREEDING BIRDS

	Falco subbuteo	Falco chicquera	Microhierax caerulescens	Spizaetus nipalensis	Spizaetus cirrhatus	Hieraaetus kienerii	Hieraaetus fasciatus	Hieraaetus pennatus	Aquila rapax	Aquila pomarina	Ictinaetus malayensis	Butastur teesa	Accipiter badius	Accipiter trivirgatus	Accipiter nisus	Accipiter virgatus	Accipiter gentilis	Spilornis cheela	Ichthyophaga ichthyaetus	Ichthyophaga nana	Pernis ptilorhyncus	Aviceda leuphotes	Leptoptilos javanicus	Ciconia episcopus	Anastomus oscitans	Ardea imperialis
Species adapted to man-modified habitats		x										x														
Species at risk in Nepal		E			V		E			R	R								E	E		E				
Species with significant breeding populations																										
Status	rwm3	r5	r4	1r2	1r5	1r5	1r4	rwm4	r5	r2	r2	r2	r4	rw2	r4	r3	r1	1r5	1r5	rm2	r3	rs2	rm3			Ex
Tropical		+	+		+	+		+	+		+	+		+	+	+		+	+	+	+	+	+	+	+	+
Tropical East Nepal			+		+	+			+		+			+				+			+	+	+	+		
Tropical Central Nepal		+	+		+	+			+		+			+				+			+	+	+	+	+	+
Tropical Far West Nepal			+		+				+		+			+				+			+	+	+	+		
Subtropical	+	+	+	+		+			+					+			+	+	+	+	+	+				+
Subtropical Schima-Castanopsis		+	+	+		+			+					+			+	+	+	+	+	+				
Subtropical West Nepal			+			+			+					+			+				+					
Lower temperate	+		+			+			+					+			+	+			+					
Lower temperate Mixed broadleaves			+						+					+			+	+								
Lower temperature Quercus lamellosa			+			+			+					+			+									
Lower temperate Quercus lanata	+		+			+			+					+			+									
Lower temperate West Nepal	+		+			+			+					+			+									
Upper temperate	+		+			+			+					+		+	+	+								
Upper temperate Mixed broadleaves			+						+					+			+									
Upper temperate Quercus semecarpifolia			+						+					+			+									
Upper temperate – West Nepal, all forest types	+		+			+			+					+		+	+	+								
Upper temperate – West Nepal, conifers only	+		+						+					+			+	+								
Subalpine			+						+					+		+	+									
Subalpine Abies spectabilis									+								.									
Subalpine Betula utilis						+																				
Subalpine Rhododendron spp.																										
Subalpine Juniperus spp.																										
Alpine																										

(continued)

Species abbreviations (column headers, left to right): T.pom = *Treron pompadora*; T.bic = *Treron bicincta*; C.ind = *Chalcophaps indica*; M.unc = *Macropygia unchall*; S.chi = *Streptopelia chinensis*; S.sen = *Streptopelia senegalensis*; S.ori = *Streptopelia orientalis*; S.tra = *Streptopelia tranquebarica*; S.dec = *Streptopelia decaocto*; C.pul = *Columba pulchricollis*; C.hod = *Columba hodgsonii*; S.rus = *Scolopax rusticola*; G.nem = *Gallinago nemoricola*; P.bic = *Porzana bicolor*; P.cri = *Pavo cristatus*; C.wal = *Catreus wallichii*; L.leu = *Lophura leucomelana*; G.gal = *Gallus gallus*; P.mac = *Pucrasia macrolopha*; T.sat = *Tragopan satyra*; I.cru = *Ithaginis cruentus*; A.ruf = *Arborophila rufogularis*; A.tor = *Arborophila torqueola*; P.asi = *Perdicula asiatica*; C.chi = *Coturnix chinensis*; C.cor = *Coturnix coromandelica*; P.hod = *Perdix hodgsoniae*; F.sev = *Falco severus*.

Category	T.pom	T.bic	C.ind	M.unc	S.chi	S.sen	S.ori	S.tra	S.dec	C.pul	C.hod	S.rus	G.nem	P.bic	P.cri	C.wal	L.leu	G.gal	P.mac	T.sat	I.cru	A.ruf	A.tor	P.asi	C.chi	C.cor	P.hod	F.sev
Species adapted to man-modified habitats					x	x			x															x	x	x		
Species at risk in Nepal	R	R		V									I			I			V				E		E	E		R
Species with significant breeding populations										+	+				*		+		+	+			+					
Status	Ir2	Ir1	Ir1	r5	r1	?s5	r1	r2	r1	r3	r3	Ir2	r5	Ex	Ir1	r?	r2	Ir1	Ir2	r4	Ir2	r2	Ex	Ir5	?s5	r?		sm5
Tropical	+	+	+		+			+	+						+			+								+		
Tropical — East Nepal	+	+	+		+										+			+										
Tropical — Central Nepal	+	+	+		+										+			+										
Tropical — Far West Nepal		+	+		+			+	+						+			+										
Subtropical		+	+	+	+	+	+		+						+			+						+	+			+
Subtropical — Schima-Castanopsis			+	+			+								+									+				
Subtropical — West Nepal			+		+	+			+						+									+				
Lower temperate				+	+		+			+	+			+			+					+						
Lower temperate — Mixed broadleaves				+	+		+			+	+						+					+						
Lower temperate — Quercus lamellosa				+	+		+			+	+						+					+						
Lower temperate — Quercus lanata					+		+										+					+						
Lower temperate — West Nepal					+	+	+		+								+											
Upper temperate							+			+	+	+					+		+	+		+	+		+			
Upper temperate — Mixed broadleaves							+			+	+						+		+				+		+			
Upper temperate — Quercus semecarpifolia							+				+						+		+						+			
Upper temperate — West Nepal, all forest types							+									+	+	+	+			+						
Upper temperate — West Nepal, conifers only																		+	+									
Subalpine							+				+	+					+	+	+	+	+							
Subalpine — Abies spectabilis																				+	+							
Subalpine — Betula utilis																					+							
Subalpine — Rhododendron spp.																				+	+							
Subalpine — Juniperus spp.																+					+							
Alpine													+														+	

(continued)

Species	Species adapted to man-modified habitats	Species at risk in Nepal	Species with significant breeding populations	Status	Tropical	Tropical East Nepal	Tropical Central Nepal	Tropical Far West Nepal	Subtropical	Subtropical Schima-Castanopsis	Subtropical West Nepal	Lower temperate	Lower temperate Mixed broadleaves	Lower temperature Quercus lamellosa	Lower temperate Quercus lanata	Lower temperate West Nepal	Upper temperate	Upper temperate Mixed broadleaves	Upper temperate Quercus semecarpifolia	Upper temperate – West Nepal, all forest types	Upper temperate – West Nepal, conifers only	Subalpine	Subalpine Abies spectabilis	Subalpine Betula utilis	Subalpine Rhododendron spp.	Subalpine Juniperus spp.	Alpine
Phodilus badius				Ex	+				+																		
Phaenicophaeus leschenaultii	x			r3	+		+		+		+																
Phaenicophaeus tristis	x			r2	+	+	+	+	+	+	+																
Eudynamys scolopacea	x			rs1	+	+	+	+	+	+	+																
Surniculus lugubris				1s1	+	+	+		+	+																	
Cuculus poliocephalus				1s2	+		+		+	+	+	+	+	+	+	+	+	+	+								
Cuculus saturatus				s1	+		+		+	+	+	+	+	+	+	+	+	+	+	+	+	+				+	+
Cuculus canorus				s1	+		+		+	+	+	+	+	+	+	+	+	+	+	+		+					
Cuculus micropterus				s1	+	+	+		+	+	+																
Cacomantis sonneratii				74	+				+																		
Cacomantis passerinus	x			s4	+	+	+	+	+	+	+																
Chrysococcyx maculatus		I		s5	+	+			+																		
Hierococcyx sparverioides		I		s2	+				+			+	+	+	+	+	+	+	+								
Hierococcyx varius	x			r1	+				+																		
Hierococcyx fugax				Ex	+				+																		
Clamator coromandus			+	1s2	+				+																		
Clamator jacobinus				s4	+	+	+		+	+	+																
Psittacula alexandri	x	R		1r2	+	+	+		+	+	+																
Psittacula cyanocephala				1r2	+	+	+		+	+	+																
Psittacula himalayana				1r1	+		+	+	+	+	+																
Psittacula krameri				1r1	+	+	+	+	+	+																	
Psittacula eupatria		E		1r5	+	+	+	+	+	+	+																
Loriculus vernalis		E		1r5	+	+	+		+	+																	
Ducula badia				1r2	+	+	+		+	+	+																
Treron sphenura		V		r5	+	+	+		+	+																	
Treron apicauda		V		r3	+	+	+		+																		
Treron phoenicoptera	x			1r5	+		+		+																		
Treron curvirostra				1r5	+		+		+																		

(continued)

Species	Species adapted to man-modified habitats	Species at risk in Nepal	Species with significant breeding populations	Status	Tropical	Tropical East Nepal	Tropical Central Nepal	Tropical Far West Nepal	Subtropical	Subtropical Schima-Castanopsis	Subtropical West Nepal	Lower temperate	Lower temperate Mixed broadleaves	Lower temperature Quercus lamellosa	Lower temperate Quercus lanata	Lower temperate West Nepal	Upper temperate	Upper temperate Mixed broadleaves	Upper temperate Quercus semecarpifolia	Upper temperate – West Nepal, all forest types	Upper temperate – West Nepal, conifers only	Subalpine	Subalpine Abies spectabilis	Subalpine Betula utilis	Subalpine Rhododendron spp.	Subalpine Juniperus spp.	Alpine
Otus bakkamoena		R		lr?	+	+	+	+	+	+	+																
Otus lempiji				lr?	+	+																					
Otus sunia		V		r2	+	+	+	+	+	+	+																
Otus spilocephalus			R	r2					+	+		+	+	+	+	+	+	+	+	+							
Bubo bubo			E	lr3	+	+	+	+	+	+																	
Bubo nipalensis			E	lr5	+	+	+	+	+	+		+	+	+	+		+	+	+								
Bubo coromandus			E	lr5	+	+	+	+																			
Ketupa zeylonensis				r3	+	+	+	+																			
Ketupa flavipes		V		lr5	+	+	+	+	+	+	+	+	+	+	+	+	+	+	+	+							
Glaucidium brodiei		E		lr2	+	+	+	+	+	+	+	+	+	+	+	+	+	+	+	+							
Glaucidium radiatum				r1	+	+	+	+	+	+	+																
Glaucidium cuculoides				r1	+	+	+	+	+	+		+	+	+	+		+	+	+								
Ninox scutulata				lr2	+	+	+	+	+	+	+																
Athene brama	×			r1	+	+	+	+	+	+	+																
Strix leptogrammica			V	lr?	+	+	+	+	+	+		+	+	+	+		+	+	+								
Strix aluco				r4							+	+	+	+	+	+	+	+	+	+	+	+	+	+	+	+	+
Caprimulgus affinis				lr2	+	+	+	+	+	+																	
Caprimulgus macrurus				r2	+	+	+	+	+	+																	
Caprimulgus indicus			R	rm2	+	+	+	+	+	+	+	+	+	+	+	+	+	+	+	+							
Zoonavena sylvatica				1/4	+	+	+	+																			
Hemiprocne coronata				lr1	+	+	+	+	+	+																	
Harpactes erythrocephalus		R		lr5	+	+	+	+	+	+		+	+	+	+		+	+	+								
Halcyon coromanda		E		lr5	+	+	+	+																			
Pelargopsis capensis		E		lr3	+	+	+	+	+	+																	
Ceyx erithacus				Ex	+	+	+	+																			
Alcedo hercules				lr5	+	+			+	+																	
Alcedo meninting		I		lr5	+	+	+	+																			
Nyctyornis athertoni		E		r4	+	+	+	+	+																		

Species	Species adapted to man-modified habitats	Species at risk in Nepal	Species with significant breeding populations	Status	Tropical	Tropical East Nepal	Tropical Central Nepal	Tropical Far West Nepal	Subtropical	Subtropical Schima-Castanopsis	Subtropical West Nepal	Lower temperate	Lower temperate Mixed broadleaves	Lower temperature Quercus lamellosa	Lower temperate Quercus lanata	Lower temperate West Nepal	Upper temperate	Upper temperate Mixed broadleaves	Upper temperate Quercus semecarpifolia	Upper temperate - West Nepal, all forest types	Upper temperate - West Nepal, conifers only	Subalpine	Subalpine Abies spectabilis	Subalpine Betula utilis	Subalpine Rhododendron spp.	Subalpine Juniperus spp.	Alpine
Merops leschenaulti	x			s2	+	+	+	+																			
Coracias benghalensis	x			r1	+	+	+	+	+	+																	
Eurystomus orientalis		R		is1	+	+	+	+	+	+	+																
Tockus birostris				r2	+		+	+																			
Aceros nipalensis		E		Ex		+																					
Anthracoceros coronatus				r2	+			+																			
Buceros bicornis		R		r4	+	+	+																				
Megalaima virens				r1					+	+	+	+	+	+	+	+	+	+	+								
Megalaima zeylanica				r1	+	+	+		+	+																	
Megalaima lineata	x			r1	+	+	+		+	+	+																
Megalaima franklinii		V		r3					+	+		+	+	+	+	+	+	+	+								
Megalaima asiatica				r1	+	+	+		+	+	+	+	+	+	+	+											
Megalaima australis				r1	+	+	+		+	+																	
Megalaima haemacephala	x			r1	+	+	+	+	+	+	+																
Indicator xanthonotus		E		r5					+	+		+	+		+	+											
Picumnus innominatus			*	r1	+	+	+		+	+		+	+		+	+											
Sasia ochracea				r4	+	+	+		+	+																	
Celeus brachyurus				r2	+	+	+		+	+																	
Picus chlorolophus		R		r4	+	+	+		+	+	+	+	+		+	+											
Picus flavinucha		R		r3	+	+	+		+	+		+	+		+	+	+	+	+								
Picus canus				r3	+	+	+		+	+	+	+	+	+	+	+	+	+	+	+	+	+	+	+	+	+	+
Picus myrmecophoneus				r2	+	+	+		+	+																		
Picus squamatus				r3	+	+	+	+	+	+	+	+	+			+	+	+	+	+	+						
Dinopium shorii				r2	+	+	+		+	+		+	+		+	+											
Dinopium benghalense				r3	+	+	+	+	+	+	+																
Chrysocolaptes lucidus				r3	+	+	+		+	+	+	+	+		+	+											
Chrysocolaptes festivus				r3	+	+	+	+	+	+	+																
Gecinulus grantia		M		r5	+	+			+	+																		

(continued)

Species	Species adapted to man-modified habitats	Species at risk in Nepal	Species with significant breeding populations	Status	Tropical	Tropical East Nepal	Tropical Central Nepal	Tropical Far West Nepal	Subtropical	Subtropical Schima-Castanopsis	Subtropical West Nepal	Lower temperate	Lower temperate Mixed broadleaves	Lower temperature *Quercus lamellosa*	Lower temperate *Quercus lanata*	Lower temperate West Nepal	Upper temperate	Upper temperate Mixed broadleaves	Upper temperate *Quercus semecarpifolia*	Upper temperate - West Nepal, all forest types	Upper temperate - West Nepal, conifers only	Subalpine	Subalpine *Abies spectabilis*	Subalpine *Betula utilis*	Subalpine Rhododendron spp.	Subalpine *Juniperus* spp.	Alpine
Blythipicus pyrrhotis		V		r5					+	+		+	+	+													
Mulleripicus pulverulentus		V		1r3	+		+	+																			
Dendrocopos himalayensis			+	r3								+				+	+			+	+						
Dendrocopos darjellensis			+	r2								+	+	+			+	+	+								
Dendrocopos cathpharius			+	r3								+	+	+			+	+	+								
Dendrocopos hyperythrus				1r2								+	+	+		+	+	+	+	+	+						
Dendrocopos mahrattensis				r3	+	+	+	+	+	+																	
Dendrocopos auriceps			+	r2								+			+	+	+										
Dendrocopos macei				r1	+	+	+	+	+	+	+	+	+	+	+	+											
Dendrocopos canicapillus				r2	+	+	+	+	+	+																	
Dendrocopos moluccensis				r3	+	+	+	+																			
Serilophus lunatus				Ex	+				+																		
Psarisomus dalhousiae		E		1r5	+	+	+	+	+	+																	
Pitta nipalensis		E	*	1r5					+	+																	
Pitta sordida		R		1s1	+		+		+	+																	
Pitta brachyura		R		1s1	+		+	+																			
Calandrella acutirostris				sm1																							+
Anthus hodgsoni				r1													+			+		+	+	+		+	
Tephrodornis pondicerianus				1r2	+	+	+	+																			
Tephrodornis gularis				1r2	+	+	+		+	+	+																
Hemipus picatus				r1	+	+	+	+	+	+	+																
Coracina melanoptera				s5	+	+	+	+																			
Coracina melaschistos	x			r3	+				+			+			+	+											
Coracina novaehollandiae	x			r1	+				+			+			+	+											
Pericrocotus flammeus	x			r1	+	+	+	+	+	+	+	+	+	+	+												
Pericrocotus brevirostris		R		r5								+	+	+													
Pericrocotus ethologus	x			r1					+	+	+	+	+	+	+	+	+	+	+	+	+						
Pericrocotus solaris		E		1r5					+	+		+	+	+													

(continued)

Appendix table (continued). Species are listed as rows; habitat / status categories as columns. A "+" indicates presence; "x" = species adapted to man-modified habitats; letters under "At risk in Nepal" give the risk category (E, R, V); "*" or "+" under "Significant breeding populations".

Species	Man-modified habitats	At risk in Nepal	Significant breeding pop.	Status	Tropical	Tropical East Nepal	Tropical Central Nepal	Tropical Far West Nepal	Subtropical	Subtropical Schima-Castanopsis	Subtropical West Nepal	Lower temperate	Lower temperate Mixed broadleaves	Lower temperate Quercus lamellosa	Lower temperate Quercus lanata	Lower temperate West Nepal	Upper temperate	Upper temperate Mixed broadleaves	Upper temperate Quercus semecarpifolia	Upper temperate – West Nepal, all forest types	Upper temperate – West Nepal, conifers only	Subalpine	Subalpine Abies spectabilis	Subalpine Betula utilis	Subalpine Rhododendron spp.	Subalpine Juniperus spp.	Alpine
Tarsiger hyperythrus			*	r3																		+		+	+	+	+
Tarsiger indicus				r3																		+	+	+	+	+	+
Tarsiger chrysaeus			+	r2																		+		+	+	+	
Tarsiger cyanurus				r1																		+	+	+			
Luscinia brunnea			+	s2													+	+	+	+	+	+	+	+			+
Luscinia pectoralis				r1												+				+	+	+	+	+		+	
Brachypteryx leucophrys		E		1r5								+	+	+			+	+	+								
Brachypteryx montana		R		R								+	+		+		+	+	+								
Brachypteryx stellata		R		1r4																					+		+
Prunella rubeculoides				r2																							+
Prunella fulvescens				r1																							+
Prunella strophiata				r2																						+	+
Troglodytes troglodytes				r2																		+	+				+
Irena puella			*	1r2	+	+																					
Chloropsis hardwickii		E		r3	+				+	+		+	+	+	+												
Chloropsis aurifrons			+	r1	+	+	+	+	+	+																	
Aegithina tiphia	x			r1	+	+	+	+	+	+	+																
Hypsipetes madagascariensis				r3	+	+	+	+	+	+																	
Hypsipetes flavalus				r2	+	+	+		+	+		+	+	+	+												
Hypsipetes mcclellandii				1r2							+	+	+	+	+	+	+	+	+	+	+						
Criniger flaveolus		V		1r2	+	+	+				+																
Pycnonotus cafer	x			r1	+	+	+	+	+		+					+											
Pycnonotus leucogenys	x			1r1					+	+	+	+	+	+		+											
Pycnonotus jocosus				1r3	+	+	+	+																			
Pycnonotus melanicterus				r3	+	+	+	+	+																		
Pycnonotus striatus			+	1r3																							
Pericrocotus roseus				1:3	+	+	+	+	+	+																	
Pericrocotus cinnamomeus	x	R		1:3	+	+	+	+																			

(continued)

	Turdus boulboul	Turdus albocinctus	Turdus unicolor	Zoothera citrina	Zoothera wardii	Zoothera marginata	Zoothera monticola	Zoothera dauma	Zoothera dixoni	Zoothera mollissima	Myiophoneus caeruleus	Monticola rufiventris	Monticola cinclorhyncha	Saxicoloides fulicata	Oenanthe deserti	Saxicola ferrea	Saxicola torquata	Cochoa viridis	Cochoa purpurea	Cinclidium frontale	Cinclidium leucurum	Hodgsonius phoenicuroides	Phoenicurus schisticeps	Phoenicurus frontalis	Phoenicurus ochruros	Phoenicurus caeruleocephalus	Copsychus malabaricus	Copsychus saularis
Species adapted to man-modified habitats					×									×	×	×	×											×
Species at risk in Nepal							R	I										E		R								
Species with significant breeding populations	+	+	+		*		*	+	+					+				*						+	+	+		+
Status	r1	r2	s3	s2	s4	?5	r4	rw2	r3	r2	r1	r2	s3	1r3	s5	rwm1	rw1	Ex	1r5	Ex	1r4	s3	r3	r1	r1	r2	r3	r1
Tropical			+											+													+	+
Tropical East Nepal			+											+													+	+
Tropical Central Nepal			+											+													+	+
Tropical Far West Nepal			+											+													+	+
Subtropical												+	+	+						+	+	+						+
Subtropical Schima-Castanopsis												+	+	+								+						+
Subtropical West Nepal												+	+	+						+								+
Lower temperate	+		+		+	+	+	+					+			+	+	+		+		+	+					+
Lower temperate Mixed broadleaves	+					+	+	+					+							+		+						+
Lower temperature Quercus lamellosa	+					+	+	+					+									+						
Lower temperate Quercus lanata			+										+															+
Lower temperate West Nepal	+		+		+								+			+	+			+								
Upper temperate	+	+				+	+						+			+	+			+						+		
Upper temperate Mixed broadleaves	+	+				+	+						+							+								
Upper temperate Quercus semecarpifolia	+	+				+	+						+							+								
Upper temperate - West Nepal, all forest types	+	+						+					+			+										+		
Upper temperate - West Nepal, conifers only		+						+					+			+										+		
Subalpine		+						+	+	+	+		+			+						+	+	+		+		
Subalpine Abies spectabilis								+	+	+			+															
Subalpine Betula utilis									+													+		+				
Subalpine Rhododendron spp.																								+				
Subalpine Juniperus spp.										+						+						+		+		+		
Alpine															+	+						+		+	+	+		

(continued)

	Seicercus xanthoschistos	Seicercus castaniceps	Seicercus affinis	Seicercus poliogenys	Seicercus burkii	Orthotomus cuculatus	Orthotomus sutorius	Prinia cinereocapilla	Prinia atrogularis	Prinia sylvatica	Prinia criniger	Prinia hodgsoni	Prinia rufescens	Prinia socialis	Prinia inornata	Bradypterus thoracicus	Cettia brunnifrons	Cettia acanthizoides	Cettia flavolivacea	Cettia major	Cettia fortipes	Cettia pallidipes	Tesia olivea	Tesia cyaniventer	Tesia castaneocoronata	Enicurus maculatus	Enicurus immaculatus	Turdus viscivorus
Species adapted to man-modified habitats							x																					
Species at risk in Nepal		I						x	x	x	x		x	x						I		R	I					
Species with significant breeding populations	+			+				*									+			*		*			+		+	
Status	r1	r3	Ex	r5	r1	Ex	r1	r3	r3	r3	r1	r2	Ex	r3	r5	r1	r5	r1	r5	r4	r2	r5	r3	r2	r1	r3	r2	
Tropical							+	+	+		+		+	+	+	+						+					+	
Tropical East Nepal							+	+					+									+					+	
Tropical Central Nepal							+	+					+									+					+	
Tropical Far West Nepal							+	+	+				+									+					+	
Subtropical	+		+				+	+		+		+	+							+		+	+			+	+	
Subtropical Schima-Castanopsis	+							+													+	+				+	+	
Subtropical West Nepal	+							+			+	+										+						
Lower temperate	+	+	+								+	+								+		+	+	+	+			+
Lower temperate Mixed broadleaves	+	+																		+		+	+	+				
Lower temperature Quercus lamellosa	+	+																		+		+	+	+				
Lower temperate Quercus lanata	+	+																					+	+				
Lower temperate West Nepal	+											+										+	+	+				+
Upper temperate	+	+		+	+															+	+	+		+	+			+
Upper temperate Mixed broadleaves	+	+		+	+																+	+		+	+			
Upper temperate Quercus semecarpifolia	+	+		+	+															+	+			+	+			
Upper temperate - West Nepal, all forest types	+				+												+	+	+		+			+	+			+
Upper temperate - West Nepal, conifers only	+				+																			+	+			+
Subalpine					+												+	+	+	+	+			+				+
Subalpine Abies spectabilis																												+
Subalpine Betula utilis																												
Subalpine Rhododendron spp.					+												+	+			+							
Subalpine Juniperus spp.																	+		+									+
Alpine																												

	Muscicapa ferruginea	Muscicapa thalassina	Muscicapella hodgsoni	Cyornis tickelliae	Cyornis banyumas	Cyornis rubeculoides	Cyornis unicolor	Cyornis poliogenys	Niltava sundara	Niltava macgrigoriae	Niltava grandis	Leptopoecile sophiae	Regulus regulus	Phylloscopus affinis	Phylloscopus fuligiventer	Phylloscopus inornatus	Phylloscopus proregulus	Phylloscopus maculipennis	Phylloscopus pulcher	Phylloscopus magnirostris	Phylloscopus trochiloides	Phylloscopus tytleri	Phylloscopus occipitalis	Phylloscopus reguloides	Abroscopus schisticeps	Abroscopus superciliaris	Abroscopus albogularis	Abroscopus hodgsoni
Species adapted to man-modified habitats																												
Species at risk in Nepal	R		V		V			E			V																E	I
Species with significant breeding populations							+							*					+	+			+					*
Status	s5	r1	r5	l:?3	?5	r3	r5	r1	r1	r2	lr4	r3	r3	rv1	r4	rvw1	r1	r2	r1	s3	svw1	?5	sm2	r1	lr2	r2	lr5	lr5
Tropical				+		+		+																			+	+
Tropical East Nepal						+		+																			+	+
Tropical Central Nepal								+																			+	
Tropical Far West Nepal					+			+																				
Subtropical	+					+	+				+																+	+
Subtropical Schima-Castanopsis	+					+	+				+																+	+
Subtropical West Nepal	+						+				+																	
Lower temperate	+	+	+	?+				+			+													+	+	+	+	+
Lower temperate Mixed broadleaves	+	+	+	?+				+			+																+	+
Lower temperature Quercus lamellosa	+	+	+	?+				+			+													+				
Lower temperate Quercus lanata	+							+			+																	
Lower temperate West Nepal	+							+			+												+	+	+			
Upper temperate	+	+	+	+				+			+					+		+	+	+	+	?+	+	+				?+
Upper temperate Mixed broadleaves	+	+	+	+				+											+	+			+	+				?+
Upper temperate Quercus semecarpifolia	+	+	+	+				+											+	+			+	+				
Upper temperate - West Nepal, all forest types		+																+	+	+	+	?+	+	+				
Upper temperate - West Nepal, conifers only																		+	+		+		+	+				
Subalpine													+	+		+		+	+				+					
Subalpine Abies spectabilis																+		+	+									
Subalpine Betula utilis																+	+	+	+									
Subalpine Rhododendron spp.																		+										
Subalpine Juniperus spp.												+				+												
Alpine												+		+														

	Spelaeornis formosus	Pnoepyga pusilla	Pnoepyga albiventer	Rimator malacoptilus	Xiphirhynchus superciliaris	Pomatorhinus ferruginosus	Pomatorhinus ruficollis	Pomatorhinus schisticeps	Pomatorhinus erythrogenys	Trichastoma abbotti	Pellorneum ruficeps	Hypothymis azurea	Terpsiphone paradisi	Rhipidura aureola	Rhipidura albicollis	Rhipidura hypoxantha	Culicicapa ceylonensis	Ficedula strophiata	Ficedula monileger	Ficedula hyperythra	Ficedula hodgsoni	Ficedula westermanni	Ficedula superciliaris	Ficedula tricolor	Ficedula sapphira	Muscicapa latirostris	Muscicapa ruficauda	Muscicapa sibirica
Species adapted to man-modified habitats					x	x	x						x					x										
Species at risk in Nepal	I			V	I						E									V				R			V	
Species with significant breeding populations			+		+			+	+									+				+		+		+		+
Status	r5	r3	r2	Ex	r5	r5	r2	r3	r1	r1	r1	r3	r3	s4	r4	r2	r1	r1	r1	r5	r3	r4	s1	r1	r5	sm4+	sm4	s2
Tropical										+	+	+	+	+	+		+	+										
Tropical East Nepal										+	+	+	+	+	+		+											
Tropical Central Nepal										+	+		+	+	+		+	+										
Tropical Far West Nepal										+	+			+	+		+											
Subtropical		+		+			+	+	+		+		+		+		+	+		+				+				+
Subtropical Schima-Castanopsis		+					+	+	+		+		+		+		+	+		+				+				+
Subtropical West Nepal								+	+		+				+		+			+				+				
Lower temperate	+	+		+			+	+							+		+		+		+	+		+	+			+
Lower temperate Mixed broadleaves	+	+													+		+		+		+	+		+	+			+
Lower temperature Quercus lamellosa	+														+		+		+		+	+		+				+
Lower temperate Quercus lanata																			+		+			+				+
Lower temperate West Nepal		+					+								+		+							+				+
Upper temperate	+	+	+	+														+	+		+		+	+	+	+	+	+
Upper temperate Mixed broadleaves		+	+		+													+	+		+		+	+	+	+		←
Upper temperate Quercus semecarpifolia		+	+		+													+	+		+		+	+				+
Upper temperate – West Nepal, all forest types			+															+	+		+	+	+	+			+	+
Upper temperate – West Nepal, conifers only			+															+	+					+				+
Subalpine			+				+											+	+				+		+			+
Subalpine Abies spectabilis			+															+	+				+		+			+
Subalpine Betula utilis			+															+	+				+					+
Subalpine Rhododendron spp.																		+										
Subalpine Juniperus spp.																												
Alpine																												

(continued)

Category	Garrulax caerulatus	Garrulax ocellatus	Garrulax rufogularis	Garrulax variegatus	Garrulax striatus	Garrulax pectoralis	Garrulax monileger	Garrulax leucolophus	Garrulax albogularis	Turdoides striatus	Turdoides malcolmi	Turdoides caudatus	Turdoides nipalensis	Paradoxornis ruficeps	Paradoxornis nipalensis	Paradoxornis fulvifrons	Paradoxornis flavirostris	Paradoxornis unicolor	Conostoma aemodium	Chrysomma sinense	Timalia pileata	Macronous gularis	Dumetia hyperythra	Stachyris nigriceps	Stachyris chrysaea	Stachyris ruficeps	Stachyris pyrrhops	Speleeornis caudatus
Species adapted to man-modified habitats										×	×	×	×							×			×					
Species at risk in Nepal	E		V		V	V								R		V	V	V								E		E
Species with significant breeding populations	+	+	+	+	+			+						*		*	*	*									+	*
Status	1r4	1r2	1r3	1r1	r3	r4	r1	r1	r1	r2	1r2	1r2	Ex	r4	1r4	Ex	1r4	r4	r2	1r1	1r1	r5	1r2	1r5	r2	r2	r2	1r4
Tropical					+	+		+	+	+		+		+								+		+				
Tropical East Nepal					+	+																	+					
Tropical Central Nepal					+	+																		+		+		
Tropical Far West Nepal																								+		+		
Subtropical		+		+	+	+	+			+	+			+												+	+	+
Subtropical Schima-Castanopsis		+		+	+	+	+																			+	+	+
Subtropical West Nepal		+				+	+		+																			+
Lower temperate	+			+		+	+							+		+									+	+	+	+
Lower temperate Mixed broadleaves	+			+		+	+									+										+	+	+
Lower temperature Quercus lamellosa	+			+		+	+									+										+	+	
Lower temperate Quercus lanata						+	+																					+
Lower temperate West Nepal		+		+		+																						+
Upper temperate	+	+		+	+			+						+	+		+	+								+		
Upper temperate Mixed broadleaves	+	+		+				+						+												+		
Upper temperate Quercus semecarpifolia	+			+				+						+														
Upper temperate - West Nepal, all forest types		+		+	+			+						+				+										
Upper temperate - West Nepal, conifers only				+				+																				
Subalpine		+		+								·				+		+										
Subalpine Abies spectabilis				+																								
Subalpine Betula utilis				+																								
Subalpine Rhododendron spp.		+		+																								
Subalpine Juniperus spp.																												
Alpine																												

(continued)

	Heterophasia capistrata	Heterophasia annectans	Alcippe nipalensis	Alcippe vinipectus	Alcippe castaneceps	Alcippe cinerea	Alcippe chrysotis	Minla ignotincta	Minla strigula	Minla cyanouroptera	Actinodura nipalensis	Actinodura egertoni	Gampsorhynchus rufulus	Pteruthius melanotis	Pteruthius xanthochloris	Pteruthius flaviscapis	Pteruthius rufiventer	Cutia nipalensis	Myzornis pyrrhoura	Leiothrix lutea	Leiothrix argentauris	Liocichla phoenicea	Garrulax erythrocephalus	Garrulax affinis	Garrulax subunicolor	Garrulax squamatus	Garrulax lineatus	Garrulax ruficollis
Species adapted to man-modified habitats																											x	
Species at risk in Nepal		E					V						E	E				E	E	R		E			V	E		V
Species with significant breeding populations		+		+	+		*	+					*	+				+		*		*	+	+	+			+
Status	r1	lr5	lr3	r1	r1	Ex	lr2	r2	r1	r2	lr5	lr5	lr3	r3	r2	lr5	lr5	lr3	r2	lr5	Ex	r1	r1	lr2	lr5	r1	lr2	lr2
Tropical		+												+						+								+
Tropical East Nepal		+												+						+								
Tropical Central Nepal		+																		+								+
Tropical Far West Nepal																				+								
Subtropical		+	+		+									+						+	+	+				+	+	+
Subtropical Schima-Castanopsis		+	+		+									+						+	+	+						+
Subtropical West Nepal			+											+						+	+							+
Lower temperate	+	+	+		+	+	+	+				+	+	+		+	+	+	+	+		+	+	+	+	+	+	+
Lower temperate Mixed broadleaves	+	+	+		+		+	+		+			+	+		+	+	+	+	+		+	+	+	+	+	+	+
Lower temperature Quercus lamellosa	+	+	+			+	+	+		+			+	+		+	+	+	+	+		+	+	+	+	+	+	+
Lower temperate Quercus lanata	+	+																	+						+			
Lower temperate West Nepal	+												+				+	+	+	+								+
Upper temperate	+			+	+		+	+	+		+					+			+			+		+	+	+	+	+
Upper temperate Mixed broadleaves	+			+	+		+	+	+							+			+			+		+	+	+		
Upper temperate Quercus semecarpifolia	+			+	+		+	+	+							+			+			+		+	+	+		
Upper temperate - West Nepal, all forest types	+			+					+							+			+						+	+		
Upper temperate - West Nepal, conifers only				+																					+			
Subalpine	+			+			+	+	+									+					+		+	+		+
Subalpine Abies spectabilis				+																					+			
Subalpine Betula utilis	+			+					+									+							+			
Subalpine Rhododendron spp.				+																				+	+			
Subalpine Juniperus spp.				+																				+	+			
Alpine																												

(continued)

Species	Adapted to man-modified habitats	At risk in Nepal	Significant breeding pops.	Status	Tropical	Tropical East Nepal	Tropical Central Nepal	Tropical Far West Nepal	Subtropical	Subtropical Schima-Castanopsis	Subtropical West Nepal	Lower temperate	Lower temperate Mixed broadleaves	Lower temperature Quercus lamellosa	Lower temperate Quercus lanata	Lower temperate West Nepal	Upper temperate	Upper temperate Mixed broadleaves	Upper temperate Quercus semecarpifolia	Upper temperate - West Nepal, all forest types	Upper temperate - West Nepal, conifers only	Subalpine	Subalpine Abies spectabilis	Subalpine Betula utilis	Subalpine Rhododendron spp.	Subalpine Juniperus spp.	Alpine
Certhia himalayana				r2								+				+	+			+	+						
Certhia discolor			+	r4								+	+	+		+	+	+	+	+	+	+			+		
Sitta cashmirensis				r1								+		+			+			+							
Sitta castanea			+	r1	+	+	+	+	+	+	+	+	+	+	+	+	+			+							
Sitta himalayensis			+	r1													+	+	+	+			+				
Sitta leucopsis			+	r2													+				+	+	+				
Sitta frontalis		E		r2	+	+	+	+	+	+		+		+	+												
Melanochlora sultanea		E		1r5																							
Parus spilonotus		E		1r4					+	+																	
Parus xanthogenys	x			r1	+	+			+	+			+														
Parus monticolus				r1	+	+	+	+	+	+	+	+	+	+	+	+											
Parus major	x			r1					+	+	+	+	+	+	+	+	+	+	+	+		+			+		
Parus ater				r2								+	+	+	+	+	+	+	+								
Parus melanolophus			+	r1	+	+	+	+	+	+	+					+	+			+	+	+	+	+		+	
Parus rubidiventris			+	r1													+			+	+	+	+	+			
Parus rufonuchalis				r2													+	+	+	+	+	+	+	+		+	
Parus dichrous			+	r3					+			+	+	+		+	+	+	+	+	+	+	+	+			
Sylviparus modestus				r2								+	+	+	+	+	+	+	+	+					+		
Aegithalos concinnus			*	r1								+				+	+			+							
Aegithalos niveogularis			+	r3							+	+				+	+	+	+	+		+	+	+			
Aegithalos iouschistos		E		r3													+										
Yuhina zantholeuca				1r2					+	+	+						+	+	+			+					
Yuhina nigrimenta		E		r5																		+					
Yuhina gularis			+	r1	+	+	+		+	+	+	+	+	+	+	+	+	+	+		+	+					
Yuhina occipitalis			+	r1	+	+		+	+	+							+	+	+			+	+	+			
Yuhina flavicollis			+	r1					+	+							+	+	+	+							
Yuhina bakeri		E		1r5					+	+										+							
Heterophasia picaoides				Ex	+				+																		

The following table lists, for each species (rows), the coded attributes and habitat/altitude occurrences shown in the original matrix. ("x" = species adapted to man-modified habitats; "E" = species at risk in Nepal; "+/*" = species with significant breeding populations.)

Species	Man-modified	At risk	Sig. breeding	Status	Tropical	Tropical East Nepal	Tropical Central Nepal	Tropical Far West Nepal	Subtropical	Subtropical Schima-Castanopsis	Subtropical West Nepal	Lower temperate	Lower temperate Mixed broadleaves	Lower temperate Quercus lamellosa	Lower temperate Quercus lanata	Lower temperate West Nepal	Upper temperate	Upper temperate Mixed broadleaves	Upper temperate Quercus semecarpifolia	Upper temperate – West Nepal, all forest types	Upper temperate – West Nepal, conifers only	Subalpine	Subalpine Abies spectabilis	Subalpine Betula utilis	Subalpine Rhododendron spp.	Subalpine Juniperus spp.	Alpine
Dicrurus leucophaeus				r1	+	+	+	+	+	+	+	+	+	+	+	+	+	+	+	+	+						
Dicrurus macrocercus	x			r1	+	+			+		+																
Lanius excubitor	x		+	1r4	+	+	+	+																		+	
Lanius tephronotus	x			r2												+	+			+		+		+		
Lanius schach	x			r1	+	+			+	+	+	+				+										
Oriolus oriolus	x			s3	+	+	+	+	+	+	+	+	+	+	+	+	+										
Oriolus xanthornus				r1	+	+	+	+	+	+	+	+	+	+	+	+	+	+	+								
Oriolus traillii				1r2	+	+	+	+	+	+	+	+	+	+	+	+	+	+	+								
Zosterops palpebrosa	x			r1	+	+	+	+	+	+	+	+	+	+	+	+											
Dicaeum cruentatum				1r5	+	+			+	+																	
Dicaeum ignipectus				r1	+	+			+	+	+	+	+	+	+		+	+	+			+					
Dicaeum concolor				1r2	+	+	+		+	+	+	+	+														
Dicaeum erythrorhynchos				r3	+	+			+																		
Dicaeum melanoxanthum		E		1r3								+	+				+										
Dicaeum chrysorrheum		E	*	1r5	+	+			+																		
Dicaeum agile				r4	+	+	+	+	+	+	+	+	+	+	+												
Arachnothera magna		E		1r3	+	+	+		+	+		+	+	+			+										
Arachnothera longirostra				1r5	+	+			+	+		+															
Aethopyga ignicauda				r2													+	+	+			+	+	+	+	
Aethopyga siparaja	x			r2	+	+	+	+	+	+	+	+	+	+	+												
Aethopyga saturata				r3	+	+	+	+	+	+	+	+	+	+	+												
Aethopyga nipalensis			+	r1	+	+	+	+	+	+	+	+	+	+	+	+	+	+	+	+	+	+	+	+			
Aethopyga gouldiae				r4													+	+	+	+	+	+	+				
Nectarinia asiatica		E		r1	+	+	+	+	+	+	+																
Anthreptes singalensis	x			1r5	+	+	+	+	+	+	+																
Cephalopyrus flammiceps			+	?4													+	+	+	+	+	+			+	
Certhia familiaris				r2													+	+	+	+	+	+			+	+
Certhia nipalensis			*	r2													+	+	+	+	+	+	+	+		

(continued)

	Lonchura punctulata	Lonchura striata	Euodice malabarica	Ploceus philippinus	Petronia xanthocollis	Passer rutilans	Gracula religiosa	Acridotheres fuscus	Sturnus pagodarum	Sturnus malabaricus	Saroglossa spiloptera	Corvus macrorhynchos	Nucifraga caryocatactes	Dendrocitta frontalis	Dendrocitta formosae	Dendrocitta vagabunda	Cissa chinensis	Urocissa erythrorhyncha	Urocissa flavirostris	Garrulus lanceolatus	Garrulus glandarius	Artamus fuscus	Dicrurus paradiseus	Dicrurus hottentottus	Dicrurus remifer	Dicrurus aeneus	Dicrurus annectans	Dicrurus caerulescens
Species adapted to man-modified habitats	x	x	x	x	x	x	x				x	x				x						x						
Species at risk in Nepal																											v	
Species with significant breeding populations											+								+	+								
Status	r2	lr3	r4	r1	r3	r2	r3	r1	t3	t2	t3	r1	r1	Ex	r1	r1	lr2	r1	r1	r3	lr3	lr1	r2	lr3	r2	s4		r2
Tropical	+	+	+	+	+		+	+	+	+	+	+			+	+	+					+	+	+	+	+	+	+
Tropical East Nepal								+	+	+	+								+				+	+	+	+	+	+
Tropical Central Nepal								+	+	+	+							+	+				+	+	+	+	+	+
Tropical Far West Nepal								+	+	+								+	+				+	+	+	+	+	+
Subtropical	+	+		+		+		+		+	+	+		+	+	+	+	+		+			+	+	+			
Subtropical Schima-Castanopsis								+								+	+	+					+	+	+			
Subtropical West Nepal						+		+			+	+	+			+		+	+		+		+	+				
Lower temperate		+			+			+				+				+		+	+	+	+	+					+	
Lower temperate Mixed broadleaves														+		+		+		+							+	
Lower temperature Quercus lamellosa														+		+		+	+	+							+	
Lower temperate Quercus lanata														+				+		+							+	
Lower temperate West Nepal						+						+				+		+	+	+	+							
Upper temperate						+						+	+					+										
Upper temperate Mixed broadleaves																		+										
Upper temperate Quercus semecarpifolia																		+										
Upper temperate - West Nepal, all forest types												+	+					+										
Upper temperate - West Nepal, conifers only													+					+										
Subalpine												+	+					+										
Subalpine Abies spectabilis													+					+										
Subalpine Betula utilis																												
Subalpine Rhododendron spp.																												
Subalpine Juniperus spp.																												
Alpine																												

	Melophus lathami	Emberiza fucata	Mycerobas carnipes	Mycerobas melanozanthos	Mycerobas affinis	Pyrrhula erythrocephala	Pyrrhula nipalensis	Pyrrhoplectes epauletta	Haematospiza sipahi	Propyrrhula subhimachala	Carpodacus rubicilloides	Carpodacus thura	Carpodacus rhodopeplus	Carpodacus edwardsii	Carpodacus vinaceus	Carpodacus rhodochrous	Carpodacus pulcherrimus	Carpodacus erythrinus	Carpodacus nipalensis	Carpodacus rubescens	Loxia curvirostra	Carduelis carduelis	Carduelis spinoides	Callacanthis burtoni	Serinus pusillus
Species adapted to man-modified habitats	×	×																							
Species at risk in Nepal									V																
Species with significant breeding populations			*	+	+			+	*	+	+	*				+	*	+					+		
Status	r2	r4	r2	r4	r2	r2	1r4	r5	1r4	1r4	r?	r2	1r2	r5	r5	r2	r1	r2	r5	r3	r4	r1	1r4	1r74	r2
Tropical	+																								
Tropical East Nepal																									
Tropical Central Nepal																									
Tropical Far West Nepal																									
Subtropical	+																								
Subtropical Schima-Castanopsis																									
Subtropical West Nepal	+																								
Lower temperate	+	+				+		+																	
Lower temperate Mixed broadleaves						+																			
Lower temperature Quercus lamellosa						+		+																	
Lower temperate Quercus lanata																									
Lower temperate West Nepal																									
Upper temperate			+	+	+	+															+	+	+	+	
Upper temperate Mixed broadleaves						+																	+		
Upper temperate Quercus semecarpifolia						+																	+		
Upper temperate - West Nepal, all forest types				+	+																		+	+	
Upper temperate - West Nepal, conifers only					+																		+	+	
Subalpine			+	+	+	+		+		+	+	+	+	+	+	+	+	+	+	+	+	+	+	+	+
Subalpine Abies spectabilis				+								+						+	+					+	
Subalpine Betula utilis					+								+				+		+				+		
Subalpine Rhododendron spp.										+					+	+	+	+	+						
Subalpine Juniperus spp.			+			+						+				+			+				+		
Alpine											+									+	+				+